ENVIRONMENTAL POLICY AND THE PURSUIT OF SUSTAINABILITY

It is increasingly apparent that human activities are not suitable for sustaining a healthy global environment. From energy development to resource extraction to use of land and water, humans are having a devastating effect on the earth's ability to sustain human societies and quality lives. Many approaches to changing the negative environmental consequences of human activities focus on one of two options, emphasizing either technological fixes or individual behavior change to reduce environmental harms through sustainable consumption habits. This book takes a different approach, focusing on the role of environmental policy in shaping the possibilities for and creating hindrances to pursuing more sustainable use of environmental resources.

This unique compilation examines environmental policy through empirical case studies, demonstrating through each particular example how environmental policies are formed, how they operate, what they do in terms of shaping behaviors and future trajectories, and how they intersect with other social dynamics such as politics, power, social norms, and social organization. By providing case studies from both the United States and Mexico, this book provides a cross-national perspective on current environmental policies and their role in creating and limiting sustainable human futures.

Organized around four key parts – Water; Land; Health and Well Being; and Resilience – and with a central theme of environmental justice and equity, this book will be of great interest to students and scholars of environmental policy and sustainability.

Chelsea Schelly is Associate Professor of Sociology in the Department of Social Sciences and the Environmental and Energy Policy graduate program at Michigan Technological University, USA.

Aparajita Banerjee recently completed her PhD in Environmental and Energy Policy from Michigan Technological University, USA.

ROUTLEDGE STUDIES IN ENVIRONMENTAL POLICY

Land and Resource Scarcity
Capitalism, Struggle and Well-being in a World without Fossil Fuels
Edited by Andreas Exner, Peter Fleissner, Lukas Kranzl and Werner Zittel

Nuclear Energy Safety and International Cooperation
Closing the World's Most Dangerous Reactors
Spencer Barrett Meredith, III

The Politics of Carbon Markets
Edited by Benjamin Stephan and Richard Lane

The Limits of the Green Economy
Matthias Lievens and Anneleen Kenis

Public Policy and Land Exchange
Choice, Law and Praxis
Giancarlo Panagia

International Arctic Petroleum Cooperation
Barents Sea Scenarios
Edited by Anatoli Bourmistrov, Frode Mellemvik, Alexei Bambulyak, Ove Gudmestad, Indra Overland and Anatoly Zolotukhin

Why REDD will Fail
Jessica L. DeShazo, Chandra Lal Pandey and Zachary A. Smith

The European Union in International Climate Change Negotiations
Stavros Afionis

The EU, US and China Tackling Climate Change
Policies and Alliances for the Anthropocene
Sophia Kalantzakos

Environmental Policy and the Pursuit of Sustainability
Edited by Chelsea Schelly and Aparajita Banerjee

ENVIRONMENTAL POLICY AND THE PURSUIT OF SUSTAINABILITY

Edited by Chelsea Schelly and Aparajita Banerjee

Routledge
Taylor & Francis Group
LONDON AND NEW YORK

earthscan
from Routledge

First published 2018
by Routledge
2 Park Square, Milton Park, Abingdon, Oxon OX14 4RN

and by Routledge
711 Third Avenue, New York, NY 10017

Routledge is an imprint of the Taylor & Francis Group, an informa business

British Library Cataloguing-in-Publication Data
A catalogue record for this book is available from the British Library

Library of Congress Cataloging-in-Publication Data
A catalog record for this book has been requested

ISBN: 978-1-138-29650-3 (hbk)
ISBN: 978-1-138-29651-0 (pbk)
ISBN: 978-1-315-09999-6 (ebk)

Typeset in Bembo
by Out of House Publishing

This one's for Prabha, who made it possible.
– CS

To Imon, my being.
– AB

This wide-ranging anthology on environmental policy does much more than the title suggests. The contributing authors present a selection of case studies that illustrate the successes and failures of environmental policies in a number of important areas including water and land use. This will be a useful supplemental textbook in courses on environmental policy.

Zachary Smith, Regents' Professor, Northern Arizona University, USA

There is no single pathway to "sustainability," but instead – as Julian Agyeman argues – "sustainabilities." The illuminating case studies collected by Schelly and Banerjee allow us to see diverse pathways in practice, to recognize the centrality of social justice, and to cultivate insights into the crucial policy contexts that both constrain and enable them.

John M. Meyer, Humboldt State University, USA; author of Engaging the Everyday: Environmental Social Criticism and the Resonance Dilemma

CONTENTS

FIGURES

TABLES

EDITORS

Aparajita Banerjee (co-editor) recently completed her PhD in Environmental and Energy Policy from Michigan Technological University, USA. She is an emerging expert in energy technology transitions, energy policy, and the social justice impacts of energy infrastructures.

Chelsea Schelly (co-editor) is Associate Professor of Sociology in the Department of Social Sciences and the Environmental and Energy Policy graduate program at Michigan Technological University, USA. She has published widely on energy technology adoption, sustainable communities, and energy policy. She is the section coordinator for the Environmental Policy and Natural Resources Management section of the Western Social Sciences Association annual conference, from which the papers in this volume are compiled.

CONTRIBUTORS

Nina Burkardt is a Social Science Research Analyst with the United States Geological Survey. She is an expert in environmental policy processes.

Joice Chang is an Assistant Professor in the Department of Politics at Humboldt State University, USA. She received a joint PhD in public policy and political science from the School of Public and Environmental Affairs at Indiana University, USA. Her current research centers on environmental voluntary agreements and legal implications of various environmental and energy policies.

Carol Considine is an Associate Professor of Engineering Technology in the Batten College of Engineering & Technology, USA. She has been active in climate resilience research since 2010 and served as the Chair of the Private Infrastructure Committee for the Hampton Roads Intergovernmental Pilot Project.

Michelle Covi is an Assistant Professor of Practice with the Virginia Sea Grant Climate Adaptation and Resilience Program at Old Dominion University, USA.

John Freemuth is Professor of Environmental Policy and the Executive Director of the Andrus Center for Public Policy at Boise State University, USA. He has written and edited three books and numerous articles on the environment and public lands, and was the chair of the Science Advisory Board of the Bureau of Land Management.

Meagan M. Jordan is an Associate Professor in the School of Public Service, Strome College of Business at Old Dominion University, USA.

M. Dawn King is a lecturer and Director of Undergraduate Studies at Brown University's Institute at Brown for Environment and Society, USA. She earned her PhD in Environmental Politics at Colorado State University, USA.

Amanda Kreuze is a PhD student in the Department of Geography, Environment, and Spatial Sciences at Michigan State University, USA. She is interested in and will continue studying the nexus between human and environmental health.

J. Gail Nicula is an Adjunct Professor in the School of Public Service, Strome College of Business at Old Dominion University, USA.

Kristin Olofsson is a PhD candidate and lecturer in public affairs at the University of Colorado Denver School of Public Affairs, USA. She is a Research Assistant with the Workshop on Policy Process Research and a Research Fellow with the Division of History and Politics at the University of Stirling, United Kingdom. Her research focuses on political behavior within the policy process and organizational determinants of participation.

Patricia Biddle Orth holds BS and MS degrees in wildlife biology and a PhD in forest sciences. She is currently a social scientist examining the linkages between the human dimensions of environmental problems, public and private land use management, and natural resource policy and governance. She pursues research that lies at the intersection of scholarly advancement and practical application within a socio-ecological context.

Erin C. Pischke is a post-doctoral researcher in Environmental and Energy Policy at Michigan Technological University, USA. She conducts energy-related research in Mexico and collaborates with others on large international Inter-American Institute for Global Change Research projects.

Beatriz Adriana Venegas Sahagún is a PhD student at the Centro de Investigaciones y Estudios Superiores en Antropología Social (CIESAS-Occidente), Mexico, and a Professor in the Department of Regional Studies – INESER, Universidad de Guadalajara, Mexico. Her research focus is environmental management at the local level. Her chapter is based on her doctoral thesis.

Roman Sidortsov, PhD, JD, LLM, currently serves as an Assistant Professor of Energy Policy at Michigan Technological University, USA. His research focuses on international and comparative energy law and policy with a special emphasis on the Russian Federation and the United States, energy security and justice, risk governance in the energy sector, and Arctic energy development.

Amelie Simons has a BS in Justice Studies from Barrett Honors College at Arizona State University, USA, and wrote this chapter while a graduate assistant in the Department of Politics and International Affairs at Northern Arizona University, USA.

Patricia Snyder is a graduate student in the Cultural and Environmental Resource Management program at Central Washington University, USA. She grew up along the banks of the Rio Grande in southern New Mexico/far west Texas and became interested in water and water policy from an early age. She currently lives in Ellensburg, Washington.

Burton St. John III is an Associate Professor in the Department of Communication & Theatre Arts, Batten College of Arts & Letters at Old Dominion University, USA.

Juita-Elena (Wie) Yusuf is an Associate Professor in the School of Public Service, Strome College of Business at Old Dominion University, USA.

INTRODUCTION

When one size does not fit all: environmental policy, social context, and social justice

Aparajita Banerjee and Chelsea Schelly

It is increasingly apparent that the currently dominant modes of production and consumption activities that define modern human societies are not suitable for sustaining a healthy global environment. From energy development to resource extraction to use of land and water, the pursuit of economic growth is having a devastating effect on the earth's ability to sustain both human and non-human lives. Even with expensive ecological degradation and damage to other aspects of the biophysical world, economic growth has failed to solve and has even in cases aggravated persistent problems like poverty, economic destitution, and crises of health and well being, causing turmoil for many people and nations around the world.

Many approaches to changing the negative anthropogenic impacts on the environment focus on one of two options. Some emphasize the options available for technological fixes, in which new technological systems are used to reduce environmental degradation or remediate impacts that have already occurred (Nordhaus and Shellenberger, 2007). Others focus on the ways individuals can change their behaviors to minimize environmental harms through sustainable consumption habits (see Shove, 2010; Heberlein, 2012). However, in both of these options, technological and behavioral, policy plays a critical role. In the former, policies can promote technological research that remediates environmental problems, helps change market mechanisms for technology diffusion, and promote user acceptance. In the latter, policies can initiate positive environmental behaviors or counteract negative ones. Furthermore, there are many options beyond these two approaches, such as changing the organization of production or focusing on structural fixes to infrastructure and planning, which also require policies. Also, environmental policies targeting technologies, behaviors, production, or infrastructure all impact the organization and daily activities of real human lives, including the experiences of human beings in terms of health and social justice. Thus, this book takes a different approach, focusing on the role of environmental policy in both shaping the possibilities for and creating hindrances to pursuing the more sustainable use of environmental resources.

Books on environmental policy often take a textbook approach, describing policy processes, formation, and evaluation in an abstract sense (Rosenbaum, 2013; Vig and Kraft, 2016). While these are helpful introductions to the basic structure of policymaking, this book is a unique compilation that examines environmental policy through empirical case studies, demonstrating with each particular example how environmental policies are formed, how they operate, what they do in terms of shaping behaviors and future trajectories, and how they intersect with other social dynamics such as politics, power, social norms, and social organization. This book also identifies some of the challenges in formulating and implementing sustainable environmental policies in terms of unintended consequences and the complexities of environmental policy across geographical scales, policy issues, and social justice. Additionally, it provides examples of possible innovation in the policy realm that can be followed to cater to new and dynamic challenges of a resource-stressed world.

By providing case studies from both Mexico and the United States, this book provides a cross-national perspective on current environmental policies and their role in creating and limiting sustainable human futures. Given the scope, the book's foundation is in policy sciences. However, the different chapters in the book represent a diverse range of disciplinary perspectives from the social sciences, including sociology, political science, and public policy, as well as interdisciplinary collaborations across environmental sciences and beyond academic institutions.

Mexico and the United States are ideally situated for examining cross-national case studies in environmental policy. As neighboring countries, both have socio-cultural impacts on one another, yet they differ widely in their socio-political and socio-economic contexts. Therefore, these countries differ in how policies are made and how they are implemented, yet they also demonstrate how similar themes of social justice and social inequality relate to environmental policymaking and implementation across national contexts. For example, in the United States, environmental policymaking often involves opportunities for public participation, which is not always the case in Mexico. Thus, many implementation challenges creep up at later stages in the context of Mexico, which affects policy success as policies are often not well-received or understood by the public. That said, formal opportunities for participation in the United States do not always provide for meaningful participation or improved policy. The use of case studies from both countries provides a more diverse exploration of the myriad challenges of environmental policymaking, and how similar problems and issues emerge even when socio-political situations are different. Besides, solving environmental problems today requires both global and local perspectives. Often, having country-specific knowledge may not be enough to understand the drivers and potential means of addressing environmental problems worldwide, as these problems manifest at local scales but are also international in scope. The use of case studies from Mexico and the United States provides exposure to policy issues that are beyond country-specific boundaries. Additionally, this book identifies policy problems, outcomes, and impacts in two different national contexts.

The unique emphasis on empirical case studies across two nations provides the context that allows this volume to advance four main arguments. The first is that environmental issues often involve complexities and potential incongruences across scale, both geographical and geopolitical. When natural resources cross the jurisdictional boundaries of localities and states, when policy choices in one location have impacts for another, and when current political boundaries do not align with the geographies of environmental problems, issues of scale mismatch arise. The second main point is that environmental policy often occurs as if divorced from other issues related to outcomes for economic, technological, social, and distributional systems, but these policy silos make for ineffective decision making and often unintended consequences, because environmental policy is always intertwined with a multitude of policy and social issues. The third argument is that environmental policy in every instance involves choices about the distribution of social justice, and that policymaking that is not explicitly attentive to social justice concerns likely works to perpetuate inequities and existing power relations rather than doing the necessary work to promote social justice via environmental sustainability. The final argument is that, given the complexities of geographical and geopolitical scale, the need to think across and beyond policy silos, and the imperative of addressing social justice issues, lacking leadership among those willing to grapple with these complexities and address current inequities is one of the biggest challenges for pursuit of sustainability via environmental policy.

Understanding sustainability

In the last couple of decades, sustainability as a single word has engendered broad interest in policymaking domains in both the public and private sectors. Since the World Commission on Environment and Development's final report, *Our Common Future*, in 1987 popularized the term sustainability and used it interchangeably with other terms like "sustainable development" and "sustainable," there have been numerous attempts to define the term and explore its core meaning (McKenzie, 2004). In what is perhaps the most popular definition, sustainable development is defined as "development that meets the needs of the present without compromising the ability of future generations to meet their own needs" (WECD, 1987, p. 9). Essentially, the very existence and quality of present and future human life depends on both the available quantity and quality of natural resources; thus, any behavior that involves the use of those resources should espouse responsibility for reducing any negative impacts of that use. Human beings are dependent on natural resource availability, and any attempt to use policy structures to manage the use of resources like energy, land, water, and even air, the resources upon which economies also depend, ought to consider the very fact that those resources are absolutely necessary for current and future human lives.

In our understanding, sustainability is a technologically enabled social construct that seeks to integrate the three popularly accepted dimensions of social (socio-political), economic, and environmental factors so that current and future development practices adhere to a plan of maintaining a fine balance between each of the dimensions. Therefore, sustainability consists of protecting amenities that have enriched life on earth as well as creating a more balanced and just world (Kemp et al., 2005). In maintaining the delicate balance, the role of humans changes from one of dominance over nature to that of a willingness to engage in adjusting their lives, activities, and societies within the context of maintaining or at least not destroying nature (Giddings et al., 2002). Existing within nature would also entail recognizing that the natural world is also home to innumerable non-human living elements and minimizing negative impacts on them. We think that at the heart of the term lies an altruistic urge of human conscience to reduce the effects of economic growth on other species and future beings of the human species.

However, it is increasingly recognized that pursuing sustainability is a Herculean effort that may require complete reconfiguration of our economic and socio-political systems (Biermann et al., 2012) and not just incremental changes in technology or human behavior. Though the task of pursuing sustainability, protecting what needs protection, and creating a more just world is daunting, it may be like many other social movements that have made modern societies possible – like the abolition of slavery, recognition of universal human rights, female suffrage, and racial and gender equality to name just a few – in that pursuing and implementing well-planned policies can get the ball rolling and provide a structure through which sustainability can be pursued in ways that are attendant to the necessary conditions of challenging existing relations of power to promote social justice and human health and well being. In this book, each chapter contributes insight into an overarching argument, which is that well-formulated policy is necessary for pursuing sustainability and that policy itself can be used to integrate socio-political, economic, and environmental considerations for the sake of a more environmentally sustainable and socially just future.

When one size does not fit all: the current context of environmental policy

The pursuit of sustainability entails protecting life-enriching facilities and services as well as creating new opportunities to provide services and amenities for advancing towards a more socially just world. Public policies play a major role in both protection of existing resources and creation of new resources, as public policies are "whatever that a government chooses to do or not to do" (Dye, 1972, p. 2). Therefore, the government can opt to make policies that work toward creating a better world or, alternately, can remain inactive in making policies that would jeopardize systems that need protection.

Policy decision makers are not alone when they make decisions. Multiple entities aid decision makers when they are making policy choices, aiming to achieve the best possible outcomes. Multiple and diverse entities are also involved in policy implementation. Participants in the policymaking and implementation process can formally include advisers coming from various backgrounds with different stakes and potential losses and benefits resulting from the policy outcome. These advisers often represent a collection of organized voices from both private and public domains putting in concerted effort to get recognition for their causes and positions. The relationship between policy makers and both formal and informal participants is often symbiotic, as policy makers depend on organized groups for making the policy work.

Therefore, policies in pursuit of sustainability may depend on the power of the policy advisers and participants to steer policy makers towards that goal. However, there are clear issues of social justice involved, apparent as soon as one starts to examine who is at the table to participate in policy formation, decisions, and implementation. One limitation to the current environmental policymaking process is that formal requirements for participation do not often gather input from the most diverse or marginalized perspectives, from those who are most likely to be negatively impacted by degradation of the natural world. Social injustices and other ethical compromises can occur if one voice among the diverse array of institutions and actors is intentionally neglected or is carried by others. The current policy context in which policy makers rely on advisers from formal and established channels can also present a challenge to efficient environmental policy in pursuit of sustainability (for example, see Jasny, Waggle and Fisher, 2015, on the "echo chambers" in current United States' climate policy networks), as existing interests and repertoires of expertise may prevent the kind of integrative thinking required to develop policy that is beneficial for all aspects of the environment (including land, water, air, ecosystems, and both the human and non-human beings that rely on these systems to maintain life), the economy, and human health as well as promoting social justice for the near and long term.

Public policies that pursue sustainability can often hit other roadblocks. These barriers can arise from the complexities that comprise the current structures and relations of power in existing socio-political domains. Political, economic, and social power can all hinder effective environmental policy. Also in this realm, different public and private institutions interact with each other and with diverse groups of people, and institutions differ in their organizational culture, mode of operation, and executive goals. Environmental policies require that diverse groups operating in diverse jurisdictions and ecological scales be able to work together, and organizational culture or incongruence between ultimate goals can present a real challenge.

The socio-political domain also consists of people and their actions, which may or may not be consistently based on their values, norms, beliefs, and aspirations (Heberlein, 2012). As individual members of civil society, people can align together by forming groups that can contest institutions and other groups. Some of the interests expressed in organized civil society may seem divorced from environmental issues, making environmental policy seem less publicly salient than other social issues. Yet as we argue in this book, environmental issues are fundamentally social issues intertwined with questions of social justice and human rights. Thus, effective environmental policy may require real attention to and real desire to remediate social injustices as well as an ability to articulate environmental policy as related to the social justice issues that are publicly salient.

Policies are also not implemented in a vacuum, and socio-cultural norms and practices that differ locally, regionally, and nationally can intersect with policy formations in ways that result in different outcomes. Therefore, policies that are successful in one part of the country can be unsuccessful in others. Besides, globalization efforts followed worldwide in the last several decades have also made decision-making susceptible to policy conditions that are often outside the control of the individual nation state, as sovereign entities are continually exposed to both positive and adverse spillovers of policy decisions from other nation states and international organizations. Thus, environmental policy that is effective in one place or at one scale may not be easily transferable, scaled up or scaled down, and sometimes the most effective environmental policy solution requires stepping outside of existing scales and jurisdictions for policy decision-making, examining international production flows or localized system needs or some combination of the local, regional, and global, rather than working at the national or state scale.

Moreover, due to the difficulties of setting fixed mandates that would be unpopular among members of the public, environmental policies can often be non-regulatory in nature and instead use the public policy tool of financial incentives. However, financial incentive policies are expensive for government coffers and difficult to support in often cash-starved governmental agencies. Additionally, non-regulatory environmental policies based on volunteerism can also depend on the whims and fancies of the stakeholder groups on whom the policies are implemented, due to a mismatch in priorities.

All this is to say, making and implementing policy is messy, as is any attempt to understand or measure policy outcomes. Policy often involves a diverse group of institutions and actors who have their own internal ambitions, objectives, or even formal policies governing behaviors. Policy also often requires an artificial attempt to isolate issues (like water, land, or air) as well as characteristics (like separating the economy from the environment from issues of social justice) and furthermore often involves ignoring how both issues and characteristics may be implicated in local cultures, regional identities, or global geopolitics. We hence advance the idea that, in environmental policymaking, one size cannot fit all, and even the idea of environmental policy as somehow divorced from issues of economics and social justice simply does not correspond to the realities of human societies.

Linking environmental policy and the thematic areas of the book

This book is a collection of 12 empirically rich case studies. The last chapter summarizes the key points and arguments advanced throughout the book and indicates how each chapter contributes to the overarching goals of the book. The chapters are divided into four thematic areas: water, land, human health and well being, and resilience. These four thematic areas encompass many of the opportunities and challenges that arise in environmental policies pursuing sustainability.

Water

One of the biggest possible threats to the existence of human and non-human life on this earth is from lack of water, due to either inappropriate use of existing water resources or inadequate access to clean water. However, water is not a bounded resource and often flows across localities, regions, and even nations. It has multiple uses and therefore involves multiple user groups belonging to different socio-political and economic backgrounds with different expectations for water resources. At times, their voices can overlap or contradict, and there may be power struggles among groups of users. On other occasions, highly effective policies with support from user groups can fall short of having the desired impacts due to the inactivity of others in the watershed region incapable of making similar policies (due to issue incongruence across geographical scale) or because of either inactivity or contradictory activities at various jurisdictional units (due to lacking coordination or support across geopolitical scales).

As a result, policies that impact water quality and water availability management are inherently complex, as multiple organized voices may arise to advise decision makers to create new policies, tweak existing policies, or refrain from any policy decisions regarding water issues. Therefore, water policies in pursuit of sustainability require policy innovation to create new policy paths or reconfigure existing setups to adjust to new challenges related to the geographical and geopolitical complexities of scale involved in managing water resources via effective public policy.

There are three chapters in this book on the theme of water, exploring the different challenges and opportunities in pursuing sustainability in environmental policies intended to address water resources. In Chapter 1, M. Dawn King takes the example of non-point source (NPS) pollution from urban runoff and nutrient leaching from agricultural soils in the Chesapeake Bay area to show that, even with policies in place, it is hard to reduce NPS pollution. One of the reasons identified is the presence of high agricultural activity using fertilizers, runoff that goes unchecked into the Bay due to the impossibility of having a centralized system for collecting agriculture runoff to check pollution loads. Moreover, many environmental policies that do incorporate NPS polluters are often dependent on volunteerism or substantial government subsidies, which do not have desired effects, and are the responsibility of small-scale contract farmers rather than the larger corporation that may have more resources available to invest in solutions. Seeking a policy solution to this problem, King proposes the use of chicken litter biochar as a fertilizer alternative to reduce NPS pollution as well as tweaking existing market mechanisms so that the benefits of biochar for both water quality and carbon emissions can be taken into account when promoting and rewarding farmers for its use.

While Chapter 1 considers water quality issues and the complexities of managing across a watershed (a theme in all three chapters on water), Chapter 2 focuses on the difficulties of developing sustainable water policies in conditions of water scarcity, particularly given conflicting ideas about and priorities for the use of water. Taking the case of water management in the Upper Klamath Basin in the American West, which experienced an economically, socially, and ecologically impactful drought in the early 2000s, Patricia Snyder identifies key stakeholders and key policy conflicts with the ability to influence water policy. This chapter demonstrates how catalyzing events can alienate groups further or bring them together, as well as showing that water-related environmental policy challenges are dynamic in nature and change depending on catalyzing events across geographical and geopolitical scales, perhaps much as the river itself changes in character as it passes through the land it travels. In her conclusion, Snyder points to the role of the federal government to act as an umbrella to bind and nurture local and regional level efforts.

Chapter 3 demonstrates that effective and sustainable water-related management policies require innovative solutions that transcend political boundaries to work on an ecosystem level where local, state, and federal entities come together to collaborate with each other and various entities of both public and private domains cohesively and coherently. In this chapter, Wie Yusuf and her colleagues present the case of Hampton Roads Intergovernmental Pilot Project, created to address adaptation needs given projected sea level rise in Southeastern Virginia, adopting a whole-of-government and whole-of-community approach in an attempt to establish a new type of institutional arrangement to solve such issues. In presenting their case, the authors point out how such projects can foster collaboration among different groups and help identify strengths. As the project was an initial step towards innovative policymaking processes to address complex environmental issues, the authors also point out significant obstacles of such an approach when policy aspirations of different actors do not align and when perceptions differ due to the ambiguous nature of the policy goals.

Land

The next thematic area focuses on land resources. Environmental policies in pursuit of sustainability related to land resources face similar issues to those of water resources, with multiple actors involved, and multiple expectations from the resource, often in contention with other actors. However, what differentiates the discussion of land and water resources to the extent covered in this book is that the benefits and the costs of land management practices are not often shared by the same group of stakeholders, unlike the water management issues discussed in the first three chapters. When it comes to land management, while benefits are diffused to wider society (benefits as wide ranging and diverse as energy development and public land access), localized impacts of the costs of a particular type of land use can become a burden to the localized population. Not only that, the policy actors playing a critical role in policies regarding land used for resource extraction may not have personal experience with the land use impacts. Additionally, members of the wider population may not be aware of or able or willing to take effective action to reduce their impacts on land resources. Moreover, issues around land are inherently environmental and economic issues simultaneously, given the economic value of natural resources that can be extracted from land resources such as underground minerals or above-ground wildlife and high-value timber. Therefore, access rights to public lands can provide significant economic benefits. The three chapters in this thematic area explain each of these issues, offering insights into the complexities and challenges of environmental policy focused on managing land.

In Chapter 4, John Freemuth takes up the case of public land management in the United States, drawing attention to its long and convoluted history, as land is held by different agencies, acquired at different times, and put to various uses, some of which often do not sit well with local level government. Some state governments have contended for ownership of public land with intentions of using that land for resource extraction that does not align well with the current multiple use policies governing federal land, which often exclude the permanent landscape level changes that mineral resource extraction entails. Moreover, the centralized aspect of federal land management has been contested, pointing out the possible benefits of local participation in resource management. As a result, multiple types of localized land management initiatives have cropped up as reformative measures, although without significant impacts. Moreover, Freemuth also points out that land acquisition under the Antiquities Act creates conflicts between local perspectives and the intended national benefits and does not include consultation with local people. All these aspects put a question mark on creating sustainable federal land management policies that would satisfy the needs of all contending stakeholders. This chapter draws attention to the long and contentious histories in managing public lands, as well as highlighting the complexities of promoting effective environmental policy across overlapping and sometimes contentious geographical scales.

In Chapter 5, Kristin Olofsson examines how different groups of policy actors participating in land use policy decisions perceive the issue. Studying policy actors involved in hydraulic fracturing, Olofsson found that most policy actors have no direct exposure to hydraulic fracturing activities, yet they perceive hydraulic fracturing as a contentious policy issue with significant potential threats to personal and national well being. Policy contention can become a critical obstacle in finding a sustainable policy solution, as policy actors may treat others who oppose their beliefs as adversaries and thus reduce the chances of developing collaborative and sustainable solutions. Perceptions among policy actors are influential in how they articulate issues and navigate policy processes, and perceptions of risk that are not based on direct experience as well as the lack of direct experience among policy actors highlight a potential social injustice when it comes to participation in land management and land use policy.

In the final chapter in this thematic area, Beatriz Venegas Sahagún draws attention to another particular land management issue: urban waste management. As Chapter 6 argues, changing consumption patterns among the least economically marginalized members of urban communities in Mexico has combined with unsustainable waste disposal behavior to increase pressure on landfill sites. In this chapter, therefore, the author stresses the significance of public participation in sustainable land use for urban waste disposal and the challenges in achieving this. The chapter also highlights the complexities of managing waste streams across multiple geopolitical scales (including federal, state, and local policies and implementation plans) and both production and consumption systems.

Human health and well being

The next thematic area of the book deals with opportunities and challenges in pursuing sustainable environmental policies related to human health and well being. As sustainability is primarily focused on human beings, human health lies at the center of it (Corvalán et al., 1999). Well being consists of material well being, quality of life, and relational well being (McGregor et al., 2015) and, arguably, a sustainable human society is inconceivable without ensuring well being for all. Human health hazards can arise from land degradation, water pollution, air pollution, energy production, industrial activities, unplanned urban waste disposal, chemical exposure, and changes in climate conditions, among many other complex socio-environmental conditions. Any perceived or actual threat from these factors can negatively impact human well being and quality of life, especially given that environmental risks are distributed unequally among social groups such that some groups experience more harm to health caused by lacking or ineffective environmental management than others.

Inequality arising from disparities in income, education, and political power, and even from factors like color and "race" and gender and sexual identity, persists in the twenty-first century. This inequality makes some groups more powerful than others, and such power imbalances create conditions of social, economic, and environmental injustice with dire consequences for human health and well being. Therefore, pursuing a sustainable society also requires addressing these inequalities in power, given the impacts they have on policies in pursuit of sustainability that challenge the current status quo of social regimes. Given this, the chapters in the thematic area of human health and well being deal with power and its manifestations that create problems in the pursuit of sustainability.

Taking the case of the Office of Civil Rights of the Environmental Protection Agency of the United States, Chapter 7 critically points out that even when policies are in place that pursue just sustainability with the intention of addressing environmental injustices, cases of environmental injustice have never been successfully addressed through the very federal office tasked with that responsibility. In this chapter, Amelie Simons take the case of the Flint water contamination crisis to show how race and economic inequality contribute to the structural inadequacies in environmental policymaking and policy implementation. In explaining the lead-contaminated water supplied to a racial minority community living in economic desperation, Simons points toward many factors that plague environmental policymaking. For example, Simons writes about how racial minority communities have historically been underrepresented in policymaking platforms which has resulted in policies mostly insensitive to their well being, and addresses how the organizational culture of the implementing agency plays a role. Therefore, this chapter illustrates that some of the challenges of achieving sustainability remain in how the existing socio-political regime operates and that overhaul of these systems may be necessary for pursuing sustainability.

Where Simons talks about power inequalities and its repercussions for federal agencies in charge of implementing environmental justice policies, Amanda Kreuze and her colleagues in Chapter 8 take up the case of power that is not highly visible (like that of a federal agency) but rather is exercised by private corporate entities as they seek to persuade private land owners of their legitimacy as land resource managers. In this chapter, the authors show how landmen in charge of acquiring access rights for hydraulic fracturing use persuasive power to coerce landowners, using tactics like misinformation; partial disclosure of information; playing on the values, beliefs, sentiments, and economic vulnerabilities of rural landowners; maneuvering discussion to mutually agreed-upon topics; and implying a legitimating connection to sovereign (state) power. Therefore, these landmen and the natural resource extraction companies they represent employ persuasive power, an indirect and insidious power that may not be highly confrontational but that is highly impactful for economically vulnerable rural landowners, to invoke support from rural communities even when such activities conflict with their health and well being.

In Chapter 9, Banerjee and Schelly shift focus to Mexico and show how municipal and village-level members of the political party in power use allocation of resources aimed to improve health and well being as a manipulative tool to promote and maintain their power. Taking the case study of a government-initiated clean cookstove dissemination program in rural Mexico, the authors show how village-level political party workers used their position of power to distribute the cookstove to community members affiliated with their party rather than to households with actual need for the technology, exemplifying a case of elite capture.

Resilience

As the goal of sustainable development is to create and maintain balanced social, economic, and ecological systems, it is increasingly being recognized that a non-linear and dynamic relationship exists between human and ecological systems (Folke et al., 2002). This leads to a complex human–nature relationship, which, in turn, requires an overhaul in the ways policies that govern how human relationships with nature are made. As changes in environmental systems or ecological regimes are no longer gradual or incremental and often happen abruptly, environmental policies in pursuit of sustainability require flexibility in the face of change, as rigidity in attempting to control environmental change through policies can yield far worse result than what they aimed to solve. In this thematic area, the chapters explore how existing environmental policies can be re-organized, re-purposed, or re-tooled in order to change in congruence with the dynamic changes in ecological systems for sustainable solutions.

In Chapter 10, Joice Chang talks about how environmental policymaking can shift from deterrence-based approaches, which currently dominate the majority of environmental policymaking processes, to cooperation-based approaches, which require a shift in the assumptions regarding the motivations and behaviors of entities regulated via environmental policy. Chang considers voluntary agreements as a potentially valuable environmental policy tool, one that is the result of a bargaining process that includes the acquiescence of the regulated entity to take on certain obligations not otherwise required by existing law and to a certain extent some compromise on the part of regulators. It is, therefore, a policy instrument that is based more on cooperation rather than deterrence. Drawing from and extending the existing social science and legal scholarships on compliance and enforcement, this chapter proposes a framework for conceptualizing the role of voluntary agreements in the hybrid model of environmental law enforcement.

In Chapter 11, Nina Burkardt and Patricia Biddle Orth consider another case of potential innovation in the tools used to develop environmental policies, especially to adapt to unknown effects of climatic change on natural resources. In this chapter, the authors describe a study to assist federal and state natural resource managers, NGOs, and citizens in developing adaptation strategies to protect key attributes ("adaptation targets") in two landscapes in southwestern Colorado. The project team included climate modelers, ecologists, and social scientists collaborating to develop a social-ecological model of the system. Throughout the three-year study, the team included key stakeholders as a means to ensure that project outputs were relevant and implementable. The study produced many products, and the final objective was to work with stakeholders to develop adaptation strategies for the adaptation targets that were identified early in the process. Chapter 11 gives an overview of the project as a whole and focuses on institutional barriers identified by on-the-ground managers, state-level managers, and managers at the federal level.

Chapter 12 takes us again to Mexico and this time Erin Pischke considers the case of Mexican oil palm production to explore how third-party sustainability certification can play a role in pursuing sustainability when government policies fail. Third-party sustainability certifications are often market-driven and held voluntarily by oil palm producers in order to access bigger markets where such certifications are mandatory, and the certifications aim to establish social, economic, and environmental sustainability across the entire supply chain process. However, both government policy and third-party certification systems face challenges in terms of ability to assist the most economically and socially marginalized members of a society.

Conclusion

In the process of compiling the chapters in this book, key themes emerged that we discuss in detail in the concluding chapter. Given the real world policy example cases used in this book, the key takeaway lessons will hopefully help readers interested in finding critical clues while contributing to environmental policies in the pursuit of sustainability. Some of the themes that run through these chapters are saddening reminders that even in the second decade of the twenty-first century, our socio-political systems are still in many cases overshadowed with social injustices, power inequalities, organizational rigidity, and overall a lack of vision that would help us transition towards a more sustainable world. However, there is also hope, and it lies in how innovations are happening in the policy realm and in how researchers are suggesting alternatives, paying attention to scale, coherence, flexibility, and participation in framing environmental policies to help us advance towards a just and socially, economically, and environmentally sustainable world.

References

Biermann, F., Abbott, K., Andresen, S., Bäckstrand, K., Bernstein, S., Betsill, M. M., Bulkeley, H., Cashore, B., Clapp, J., Folke, C., and Gupta, A., 2012. Navigating the Anthropocene: improving earth system governance. *Science, 335*(6074), pp. 1306–1307.

Corvalán, C. F., Kjellstrom, T., and Smith, K. R., 1999. Health, environment and sustainable development: identifying links and indicators to promote action. *Epidemiology-Baltimore, 10*(5), p. 656.

Dye, T. R., 1972. *Understanding Public Policy*. Englewood Cliffs: Prentice-Hall.

Folke, C., Carpenter, S., Elmqvist, T., Gunderson, L., Holling, C. S., and Walker, B., 2002. Resilience and sustainable development: building adaptive capacity in a world of transformations. *AMBIO: A journal of the human environment, 31*(5), pp. 437–440.

Giddings, B., Hopwood, B., and O'Brien, G., 2002. Environment, economy and society: fitting them together into sustainable development. *Sustainable Development, 10*(4), pp. 187–196.

Heberlein, T. 2012. *Navigating Environmental Attitudes*. New York: Oxford University Press.

Jasny, L., Waggle, J., and Fisher, D. R., 2015. An empirical examination of echo chambers in US climate policy networks. *Nature Climate Change, 5*(8), pp. 782–786.

Kemp, R., Parto, S., and Gibson, R. B., 2005. Governance for sustainable development: moving from theory to practice. *International Journal of Sustainable Development, 8*(1–2), pp. 12–30.

McGregor, A., Coulthard, S., and Camfield, L., 2015. Measuring what matters: The role of well-being methods in development policy and practice. Working Paper. Overseas Development Institute [online]. Available at: nrl.northumbria.ac.uk/id/eprint/23421.

McKenzie, S., 2004. Social sustainability: towards some definitions. Hawke Research Institute Working Paper Series No. 27 [online]. Available at: w3.unisa.edu.au/hawkeinstitute/publications/downloads/wp27.pdf [Accessed 10 August 2017].

Nordhaus, T. and Shellenberger, M., 2007. *Break Through: From the Death of Environmentalism to the Politics of Possibility*. Houghton Mifflin Harcourt.

Rosenbaum, W. A., 2013. *Environmental Politics and Policy*. CQ Press.

Shove, E., 2010. Beyond the ABC: climate change policy and theories of social change. *Environment and Planning A*, 42, pp. 1273–1285.

Vig, N. J. and Kraft, M. E., eds., 2016. *Environmental Policy: New Directions for the Twenty-First Century*, Ninth Edition. CQ Press.

World Commission on Environment and Development (WCED), 1987. *Our Common Future*. Oxford: Oxford University Press.

PART I

Water, water management, and adaptation to changing water landscapes

1

NUTRIENT TRADING CREDITS

Best management practices and policy hurdles for non-point source polluters in the Chesapeake Bay

M. Dawn King

Since the 1960s, the United States has made large strides in cleaning up waterways and air quality through both regulatory and market-based policy tools, but serious problems remain. One policy hurdle to cleaning up waterways is that water pollution is an interstate problem – pollutants created in one state, often to the economic benefit of the polluter, can cause environmental and economic damage in other states and waterways downstream. Given most environmental policy implementation occurs at the state level, multi-state agreements are often difficult to manage and implement without federal oversight (Rabe, 2016). In the case of the Chesapeake Bay, it took almost three decades from the initial formation of the Chesapeake Bay Agreement to the implementation of legally binding pollution limits (Hoagland and Mueller, 2011).

Another large policy hurdle to increased air and water quality is how to deal with diffuse non-point source (NPS) polluters. Most pollution control success to date is due to targeting point source (PS) pollution – where the discharge occurs at one location. These point sources, such as power generators, factories, and wastewater treatment plants, are much easier to pinpoint, monitor, and regulate than NPS polluters, such as automobiles or agricultural runoff. In the Chesapeake Bay, NPS forms of pollution, such as urban runoff and nutrient leaching from agricultural soils, are diffuse and incredibly difficult to quantify per site. This is particularly problematic with NPS water pollution from the agricultural sector, as the estimated 83,000 farms in the region (USDA, 2017) account for an estimated 42% of nitrogen (N) and 54% of phosphorus (P) loads to the Chesapeake Bay area, largely through excess fertilizers, herbicides, and insecticide use, while wastewater treatment plants only account for around 20% of effluence (Van Houtven et al., 2012).

There are policy tools in place to include NPS polluters in water and air quality plans, but many are not successful at decreasing NPS pollution in practice due to high transaction costs, farmers' hesitance in joining government programs, and difficulty in monitoring and measuring pollution reduction on farms (Fisher-Vanden and Olmstead, 2013; Ribaudo and Gottlieb, 2011). Further, many environmental policies that do incorporate NPS polluters are often dependent on volunteerism or substantial government subsidies, since the federal government does not have jurisdiction to regulate most farms under the Clean Water Act (CWA). One way to potentially sidestep the federal government's lack of regulatory authority on farm pollution, especially with nutrient leaching into soils, is through policy tools that reduce farm pollution through market mechanisms as a way of encouraging farmers to participate. This chapter explores the viability of one such policy tool, nutrient trading credits, as a means to incorporate NPS polluters into the Chesapeake Agreement. It also evaluates possible policy obstacles to using poultry char, the by-product of heating chicken litter in the absence of oxygen, as an approved on-farm best management practice (BMP) to include in nutrient trading schemes. While the Chesapeake Bay Agreement established acceptable NPS BMPs, such as reduced tillage and land conservation, to incorporate farmers into water quality trading schemes, there may be a need to incorporate more BMPs given farmers' lack of interest in joining trading schemes to date. However, agreement on technology-driven solutions to reduce NPS pollution, such as manure treatment, are often difficult to incorporate into acceptable BMPs as all new BMPs must pass through a rigorous bureaucratic and scientific evaluation process and most current studies are small-scale, and geography and weather impact soil amendments' effectiveness.

Through analyzing state nutrient trading policies, subcommittee reports to the Chesapeake Bay Program, and current scientific reports on chicken litter biochar, this research finds that the use of poultry char as a soil amendment may reduce nutrient loads in the Bay, and more states are investigating the role of manure treatment technology as a possible solution to NPS pollution. However, personal interviews conducted with actors within each of the Chesapeake Bay states' nutrient trading oversight agency, as well as actors within the Environmental Protection Agency (EPA), the United States Department of Agriculture (USDA), and private organizations working on the Chesapeake Bay Program, suggest research will need to quantify biochar's effectiveness before policymakers will adopt this technology as a BMP. Even though there are transactional, socio-psychological, and measurability problems associated with incorporating farmers into pollution-reduction schemes, it is important to keep advancing research on technology-driven BMPs. Further, policymakers could look toward manure treatment technologies such as biochar production as a viable tool not only in nutrient trading schemes but also in regional greenhouse gas (GHG) trading schemes. Technological innovation must be coupled with governance innovation, such as incorporating agriculture into both nutrient and carbon credit trading schemes, when trying to reduce pollution in a region.

The Chesapeake Bay and nutrient trading credits

The Chesapeake Bay Watershed encompasses parts of New York, Pennsylvania, Delaware, Maryland, Virginia, and West Virginia. The region is particularly prone to nutrient runoff, as the watershed is 64,000 square miles and averages only 21 feet in depth (Van Houtven et al., 2012). For decades, scientists have raised concerns regarding ecosystem health in the region, as excessive amounts of N and P runoff in the Bay fuel the growth of harmful algal blooms and sediments that can block sunlight from underwater plants, causing ecological harm to the Bay while negatively impacting the seafood and tourist industries. The sources of nutrient pollutants include urban and agricultural runoff, wastewater treatment facilities, and industrial facilities. While individual states worked together as early as the 1980s to help cooperatively solve the water pollution problem in the region, it took a federal intervention to force states to follow through on their implementation plans.

The original Chesapeake Bay Agreement, signed in 1983 by Virginia, Maryland, Pennsylvania, the District of Columbia, the EPA, and the Chesapeake Bay Commission, was a written acknowledgment that cooperation is necessary to clean up the Bay. This was followed by a 1987 agreement establishing, among other things, a numeric goal of a 40% reduction in both N and P loads by 2000. Later, the Chesapeake 2000 Agreement brought in the headwater states of New York, Delaware, and West Virginia to guide restoration efforts for the following decade (Chesapeake Bay Program, 2017). While regional cooperation is key to solving many environmental problems, the state-led agreements faced complications similar to many interstate environmental agreements – lack of funding, regulatory oversight, and enforceability (Rabe, 2016; Weibust, 2009). After years of state-led attempts to reduce pollution fell short of achieving goals, the federal government became a more active partner to mandate and support implementation plans.

Under federal law via the CWA's mandate to ensure all waterways are "fishable and swimmable," President Obama signed Executive Order 13508 on 12 May 2009, to ensure Chesapeake Bay protection and restoration through a federal and state partnership (Executive Order 13508, 2009). The order mandated a pollution diet and set pollution goals for the Bay via EPA regulatory authority to monitor and enforce state efforts to reach total maximum daily loads (TMDLs) of N, P, and sediment. Fortunately, the executive order provided funding and oversight to strengthen partners from multiple state and federal agencies, local governments, non-profit organizations, and academic institutions. The order established a federal leadership committee (the Committee), led by an EPA administrator and composed of senior officials from multiple departments including Agriculture, Commerce, Defense, Homeland Security, Interior, and others to work with states to oversee the development and coordination of programs. The Chesapeake Bay Agreement is now a model of interstate environmental cooperation and is making progress in achieving the 2025 TMDL goals.

The Chesapeake TMDL established pollution control measures to be in place by 2025 in which 60% of actions must be completed by 2017. The EPA used modeling to determine TMDLs for 92 specific Bay tidal segments, with implementation strategies to achieve TMDL goals drafted and implemented by individual states (EPA, 2010, ES-5) in the form of watershed implementation plans (WIPs). These plans vary by state but include storm water and wastewater, permitting stricter controls on agriculture, funding for manure technology, and imposing strict limits on wastewater treatment plants (EPA, 2010, ES-1–2).

Nutrient trading credits and state implementation

States have historically managed NPS polluters, but the federal government has authority via the CWA to regulate point sources of water pollution, namely wastewater treatment facilities, with the National Pollution Discharge Elimination System (NPDES). The NPDES requires point sources of water pollution to purchase permits before discharging into surface water and often requires plants to abide by tech-based effluent limits. However, after the establishment of a TMDL, some state WIPs allow new wastewater treatment facilities to purchase pollution offsets from NPS polluters, such as farms who agree to decrease their on-farm discharges via nutrient management strategies, conservation tillage, land retirement, and other BMPs (Ribaudo and Gottlieb, 2011). These offsets could be a powerful market mechanism to reduce overall nutrient load in the Bay, given close to half of the pollution stems from agricultural runoff. In theory, PS to NPS nutrient trading makes sense, and economic reports on trading simulations posit the inclusion of NPS polluters into the trading scheme could be significantly less costly than not including NPSs (Van Houtven et al., 2012). However, the complications of measuring, managing, and monitoring NPS pollution has led to few actual PS to NPS trades.

After interviewing over 20 state and national experts charged with monitoring and implementing TMDLs in the Bay, one point of agreement is that incorporating NPS polluters into trading schemes creates a logistical nightmare. Much of the literature on nutrient trading supports their claims. First, trading programs often offer farmers few economic incentives, as it is difficult for many farms to reach the baseline level of compliance, let alone go above and beyond the baseline to start generating trading credits. However, these baselines are crucial to ensure farmers do not misrepresent their initial management strategies. For example, a farmer who is currently implementing nutrient management practices would have the incentive to lie to regulators about these practices in the hope of claiming they are "new" practices to earn credits. If a farmer receives credits for practices already in place, or "non-additional credits", pollution in the Bay may increase. A baseline encourages farmers to claim their current best management practices meet that level, which ensures fewer non-additional credits enter the market (Ribaudo and Savage, 2014). While farmer implementation of nutrient management practices may be a particularly efficient mechanism to reduce load, as it also decreases farmer expenditures on inputs, it is also especially prone to supplying non-additional credits, since regulators must trust the farmer's claim that the practices are new. This is different from BMPs such as conserving land or creating buffer zones that are much easier for regulators to identify.

Second, as stated above, NPS leaching is difficult to monitor and measure, which leads to many extra hurdles when trading NPS credits, including higher compliance costs for states to supervise diffuse actors. Some states implement a trading ratio of

2:1 or 3:1 to offset the lack of exact metrics for farm nutrient leaching. However, this forces PS polluters to purchase at least double the credits from farms, and many interviewees who handle state implementation claim this makes it an unpopular option.

Third, when farmers do enter the trading market, the voluntary approach to NPS pollution control leads them to choose BMPs that make the most economic sense for themselves instead of BMPs that may contribute the most to the public interests; in other words, they are supply driven rather than demand driven (Shortle et al., 2012). Fourth, the application process to enter the trading scheme can be arduous and daunting for farmers; one USDA official interviewed claimed some applications are 50 pages or longer depending on the chosen BMPs. Fifth, since the EPA does not currently regulate farmers, they may be reluctant to enter markets, as they may interpret the oversight as a form of regulation (Fisher-Vanden and Olmstead, 2013; Ribaudo and Gottlieb, 2011). Finally, all interviewees, from scientists to regulators, felt the success rates for on-farm BMPs are inconsistent and difficult to measure due to unpredictable variables such as precipitation and weather, and many argued there is simply no incentive for farmers to participate in nutrient trading programs without subsidized or cost-sharing programs since the federal government does not regulate NPS polluters.

Given the many downsides of incorporating NPS polluters into nutrient trading schemes, the solution to decreasing agriculture nutrient leaching through a trading scheme may rest in either reducing allowable BMPs to those that can best quantify leaching reductions, increasing the types of allowable BMPs available for farmers through utilizing new technology, or changing the policy approach altogether by incorporating biochar into the energy cap and trade market. Virginia has used the first strategy, successfully keeping transaction costs low by limiting eligible BMPs to land conversion only (DeBoe and Stephenson, 2016). This is easy to monitor, much like the introduction of cover crops, and almost guarantees less nutrient use. However, this limits the many other ways farmers may reduce nutrient loads, such as better nutrient management practices that not only reduce nutrient load but also save farmers' money on inputs.

Another option is to try increasing the allowable BMPs to give more flexibility to farmers and to take advantage of newer technologies, such as remote sensing, computer modeling, or manure technologies, that may make nutrient reduction more quantifiable. Currently, states allowing multiple avenues to receive credits still produce few nutrient trades involving NPS polluters, likely given the transaction and monitoring costs mentioned above. However, if researchers can better quantify the impact of biochar processing, or manure treatment technologies in general, it may provide a tool to bypass the monitoring, trade ratio, baseline, and non-additionality problems mentioned above. It is important to note, however, that I am not arguing biochar processing *should* be included as a BMP, as we need more research on biochar's impact on nutrient leaching and lifecycle emissions, but, rather, I use it as a case study to identify policy hurdles to integrating new technologies in environmental agreements.

Biochar and its potential role in decreasing nutrient leaching

While Pre-Colombian Amazonians used biochar as a soil amendment as early as 2,000 years ago, contemporary science is just starting to investigate and measure the exact mechanisms that make biochar such a valuable soil amendment. The simple act of heating biomass in the absence of oxygen (pyrolysis) creates a charcoal substance that can potentially sequester carbon, increase crop yields, and minimize leaching of nutrients into waterways – depending on the biochar input and soil type. Most commonly, biochar is derived from plant matter, and research is mixed on its effectiveness. However, recent research investigating the impact of poultry char on crop production and air and water pollution seems more promising. Poultry char is unique in that it is dense in nutrients that become fixed with the carbon, and research findings suggest that poultry char as a soil amendment may reverse N leaching and decrease nitrogen oxide (N_2O) and methane (CH_4) emissions while simultaneously increasing crop yields (Cowie et al., 2015; Woolf et al., 2010). In addition, the saturation of chicken farms in the Chesapeake Bay region creates nutrient "hot spots" near chicken facilities, since poultry farms often give away raw manure or offer it at a very low cost to local farms. For the farms, this becomes a very cost-effective way to fertilize soils, but they also tend to overuse raw manure, which causes excess N and P leaching into waterways while releasing GHGs into the atmosphere. If poultry char application is permittable as a BMP, trading credits could help offset the costs of purchasing biochar as a soil amendment for farmers accustomed to receiving free fertilizer. Further, recent data suggests that the use of chicken litter as the biochar input could increase crop yields, becoming yet another incentive for a farmer to rotate poultry char into their BMPs (Jeffery et al., 2015). However, the political process of accepting a BMP under the Chesapeake Model is lengthy and arduous, given that research must demonstrate its impact at the farm level.

In order for a BMP to count toward a state's required TMDL reduction for the Bay, it must be approved for the Chesapeake Bay Model with EPA oversight. This is no easy process, as the EPA has strict guidelines requiring scientifically sound research proving reduced leaching. In 2016, the Chesapeake Bay Program Manure Treatment Technologies Expert Panel (Hamilton et al., 2016) followed up on a 2014 study with an in-depth review of manure treatment strategies. The expert group finally recommended that manure treatment technologies be added to the next phase of modeling tools, which is dependent on Chesapeake Bay Program approval. Even then, states would have to verify and document the efficacy of the manure treatment strategies in decreasing leaching, which is harder to quantify than increased crop yields. Scientists are currently investigating biochar's quantifiable impact on nutrient leaching in the Bay, but the Chesapeake Bay Program will need to see more published data before adding it as a BMP. However, beyond the potential to serve as a soil amendment for farmers interested in decreasing nutrient loads, biochar processing could become an engine of economic growth for large-scale biochar processors and/or large chicken processing companies, as processors could potentially reduce CH_4 emission through better manure handling strategies, which may allow them to enter carbon trading schemes much more quickly than farmers can enter the nutrient trading market.

Biochar potential to enter carbon offset markets through processors

The obstacles prohibiting most farmers from entering nutrient trading markets may mean that, in addition to increasing BMP options, a completely separate policy approach could be necessary to decrease NPS pollution caused by the overuse and improper handling of raw manure. Water laws and nutrient trading schemes may just need more time to develop, most likely with the help of federal intervention and federal spending on increasing capacity to measure and monitor BMP impacts. This is especially the case in the Chesapeake Bay and similar agricultural watersheds, as the EPA has stepped in specifically to deal with water leaching and pollution. However, as with many environmental problems, policy solutions often have to be interlinked. Many of the same agricultural practices that contribute to nutrient leaching also contribute to GHG emissions, and regional carbon markets are already established trading schemes that have the potential to incorporate agriculture into carbon offset markets. In general, carbon offset markets allow GHG emitters to offset their emissions through purchasing credits elsewhere. Similarly, some states allow manure treatment facilities that capture syngas, or synthesis gas, for energy use to qualify for renewable energy certificates (RECs) – payments for producing renewable energy.

Worldwide, the agriculture sector is responsible for around 15% of GHG emissions (EPA Global, 2016), not including the impacts of land use change often tied to increased agricultural production. Agricultural practices and soil management in the United States account for almost 80% of total nitrous oxide (N_2O) emissions, which is 300 times more potent than carbon dioxide (CO_2) (EPA Nitrous, 2016). Handling and composting of raw manure, paired with excess N leaching into waterways, leads to N_2O emissions.

When soil N leaches into groundwater, nitrification and denitrification occur, which can lead to N_2O emissions; thus, any reduction in nutrient leaching could reduce GHG emissions. Many of the same BMPs listed for nutrient trading qualification also reduce GHGs, and biochar application is often mentioned in GHG reduction strategies, given its potential ability to store carbon.

Agriculture and soil management practices release potent GHGs into the air, but agriculture can also act as a means to store carbon. Carbon sequestration refers to the long-term retention of carbon in plant biomass or soil. The nutrients in biochar are water insoluble, making them less likely to leach, and the carbon is fixed, which provides a slow releasing food source for soil microbes – keeping carbon trapped in the soil instead of being released to the atmosphere. Concurrently, the increase in microbial activity breaks down the nutrients that plants need to grow. Since plants store CO_2 in their biomass, any increase in plant production leads to more carbon storage. While better manure and soil management practices could lead to GHG offsets, biochar application could potentially be the BMP best equipped to increase carbon storage in soils (Woolf et al., 2010).

The Regional Greenhouse Gas Initiative (RGGI) is the first regional market-based regulatory program in the United States to reduce GHG emissions and does allow for agricultural GHG offsets. Starting in 2007, it is an auction-based system in which fossil-fuel power plants purchase emission allowances, the emission cap is reduced every year, and the auction money is allocated to states to reinvest in efficiency, renewable energy, and consumer projects. It is now a cooperative effort among Connecticut, Delaware, Maine, Maryland, Massachusetts, New Hampshire, New York, Rhode Island, and Vermont. To date, it has been successful in reducing emissions from power plants in the region while simultaneously keeping electricity prices stable (Carson, 2015). While offsets are not common, RGGI regulations allow power generators to offset 3.3% of their CO_2 compliance obligation (RGGI, 2016). One offset is linked to avoiding CH_4 emissions through agriculture manure management, but it's currently only for anaerobic digestion of manure. Additionally, some states allow gas created through anaerobic digestion of manure to qualify for RECs. Similarly, the California Air Resources Board approved BMPs in rice cultivation for CH_4 offsets under its cap and trade program in 2015. While these offsets are minute compared to caps placed on power generators, it does demonstrate policymakers are steadily looking at means to decrease agricultural GHG emissions as part of a larger mitigation strategy, and, once quantifiable, carbon capture on farms could potentially become a large market.

Currently, the Chesapeake Bay area does not have a carbon trading market, but the Obama administration suggested (Federal Register, 2015, section VIII) regional or in-state cap and trade policies as a means to abide by the 2015 Clean Power Plan (CPP). If the Plan continues under the current administration, or if regions decide to push ahead regardless of CPP implementation, states will choose policy tools or market mechanisms to reduce GHG emissions from power plants. Delaware, Maryland, and New York are already part of RGGI, and poultry producers in these states could investigate the steps necessary to start receiving offset credits for CH_4 reductions through manure treatment practices. Since current GHG reduction policies do not legally require the agricultural sector to participate, adding manure treatment technologies as possible offsets could be one way to encourage farmers to voluntarily reduce on-farm emissions. Further, pyrolyzing raw manure offers many benefits, which may also encourage farmers to enter the carbon trading market. First, pyrolysis reduces the volume and weight of manure, making it easier to transport. This is particularly beneficial for farms in the Chesapeake region, as it can allow them to more efficiently export biochar as a soil amendment both within and outside the region. Second, feedstock costs are a large prohibitive factor in the commercialization of plant-based biochar facilities over the lifetime of the plant (Shackley et al., 2015), but manure separates itself from plant-based or cellulosic inputs. At present, manure is regulated with regards to handling and storage, and farmers may already have to pay to dispose of the waste, unlike forest thinnings that are expensive to aggregate, agricultural waste that can be left in the field for free, or woodchips that could be in high demand for other uses. If feedstock sources, pyrolysis units, and consumer markets are near each other (as Roberts et al., 2009 suggest), manure could become a cheap input with a consistent supply for poultry char producers. Third, pyrolysis plants can produce more than just biochar. Excess gases from the burning process or recovered heat can be trapped for heat or electricity to offset fossil fuel use in the production process.

According to one interviewee, a poultry farmer in the Bay who purchased a biochar gasifier with help from a Conservation Innovation Grant and his state's Department of Agriculture, his 800,000 birds can produce 125 to 600 tons of wet poultry litter per year, or 25 to 120 tons of biochar. He sold his first ton at a net price of $480 to an organic farm in New Jersey, compared to $5/ton for wet litter. In addition, he used the excess heat from the process to help heat his poultry houses, saving him about 4,000 gallons of propane. This also lowered the humidity in the house, which he claims led to a 7% increase in bird size. However, the initial cost of the gasifier was close to $1 million, and upkeep and maintenance can be difficult. He is also running a pilot program, so many are waiting to see if his biochar model works. The growth of on-site biochar gasifiers will most likely need government subsidies to expand, but, eventually, if the market begins to grow, an additional income stream for producing char, such as credits from a carbon offset market, may be enough incentive to encourage poultry farmers to produce biochar without subsidies.

Policy opportunities to encourage processors to enter carbon trading schemes

Since commercialization of the biochar market is relatively new, there are policy tools that could help develop the market. Certainly, state and federal subsidies to support biochar production facilities could help. Likewise, Conservation Innovation Grants through the USDA can supply much needed start-up capital for biochar facilities. Some recommend allowing farmers to combine USDA subsidies they already receive for environmental services, such as land conservation, with trading credits. Currently farmers cannot "double dip" incentive programs (Fisher-Vanden and Olmstead, 2013). Further, federal and state governments can expand research into the impacts of manure treatment strategies on reducing GHGs and actively seek large poultry or other livestock farmers to enter the market. States can also push regulatory best management practices, impose taxes, or offer RECs or offset credits to large-scale processors to encourage technological development in manure management.

Currently, contracted farmers, not large-scale processors, are responsible for raising chickens and waste removal. Spreading the litter onto nearby fields is the cheapest option for waste removal, so it is the most common mechanism used. Maryland is one of the few states that introduced legislation to make agri-business help with cleanup costs. Legislators introduced a bill that would force large companies to pay a fee of 5 cents per bird (called the Poultry Fair Share Act) that would bring in close to

$20 million to help fund farmers to adopt BMPs. In an article written in 2014, John Upton points out Maryland taxpayers contribute $110 million to a Bay Restoration Fund, while the large-scale poultry corporation Perdue pays nothing (Upton, 2014). While the Maryland governor threatened to veto the bill before it even received a hearing, the attempt to introduce such legislation demonstrates that states may focus on poultry companies in the future to pay for their nutrient pollution.

States can also include manure treatment technologies in their renewable energy goals *and* push for nutrient trading as a strategy to reduce the state's nutrient leaching. In Pennsylvania, the Gettysburg Energy and Nutrient Recovery Facility not only receives N and P credits, accounting for 1.59% and 0.55% respectively of Chesapeake Bay-wide 2025 reduction goals, but also receives Pennsylvania RECs for their on-site gasification (EnergyWorks, 2017). Since manure management and gas generation is at a poultry farm, the company can measure reductions of N and P, which gives them a 1:1 trading ratio in the market. The chicken farm agreed to host the facility as the gasification plant takes ownership of the manure once it is in storage barns, pays the farm consistently per ton of manure, and offers the farm consistent, reduced electricity prices. The partnership benefits both the gasification facility and the farm. One interviewee with a bioenergy company insists that Pennsylvania's focus on nutrient credits in their state water implementation plan has led to manure treatment technology advancements.

Conclusion

This chapter demonstrates many difficulties associated with environmental governance. First, environmental pollution often crosses state borders, making it an interstate problem, where those polluting may not be those who environmentally and economically suffer from the pollution. Since policies in the United States are at the state or federal level, federal intervention is often needed with interstate conflicts. Second, federalism makes NPS pollution particularly difficult to regulate, as the federal government only regulates PS polluters, making many NPS programs voluntary to farmers and limiting innovation to reduce pollution at the farm-level. Third, while many agree that technological solutions to pollution control will be necessary, quantifying a new technology's impact is costly and lengthy. This is especially relevant when there is a lack of lifecycle analysis for new technologies – such as carbon emissions produced during pyrolysis. While many studies focused on chicken litter biochar show promise in its ability to store carbon and limit runoff (Cowie et al., 2015; Woolf et al., 2010), decision makers want to see quantifiable reductions in pollution before adding any new technology as a BMP. In politics, it is also important to have decision makers who are impartial and scientifically literate. This is often difficult to find on voluntary panels, especially when certain industries may try to push their own technologies over others. As one interviewee who worked on the Chesapeake Bay Program agriculture work group noted, it's not just about sound science, but also "ensuring that the expert panel process is conducted with the expertise and science background necessary" to make proper policy decisions. Finally, pollution abatement policies often operate as silos focused on water, air, or waste issues when environmental problems are often interlinked. While there are many policy obstacles to regulating interstate NPS water pollution, this chapter also finds that regional and state governance bodies are working to better integrate NPS polluters through innovative policy solutions.

The potential to include manure treatment technologies, specifically biochar, in both climate and nutrient trading schemes could be one way to encourage farmers to adopt more BMPs that will reduce both GHGs and nutrient leaching. Any major push to further clean up waterways must include reductions in agricultural nutrient leaching and must go beyond traditional state-level policy. Certainly, nutrient trading schemes are not the only way to reduce NPS pollution. Many scholars point toward farmer education programs and increasing government subsidies for conservation programs to reduce NPS pollution. However, given the lack of regulatory oversight, market mechanisms may also be a viable tool to encourage farmers to enter nutrient trading markets. Initial research on chicken biochar demonstrates that it could potentially serve as one such additional BMP for farmers, but biochar's role in reducing GHGs on-site could be much easier to quantify in the carbon market while also providing NPS pollution reduction benefits.

Manure management processes that lead to CH_4 reduction are already part of RGGI's carbon trading programs and some states currently allow the biogas by-product of manure treatment (predominantly CH_4) from anaerobic digestion to qualify for RECs, so research on pyrolysis's role in decreasing CH_4 could provide support to add biochar producers into GHG trading schemes as well. Paired with the potential economic benefits that come with biochar production facilities, entrance into carbon markets may be enough to spur growth in the sector, which would lead to more biochar on the market for farmers to use as a BMP. Further, potential regulatory policies holding large poultry processors liable for pollution may also spur growth in biochar research.

If incentives and regulations spur growth in the biochar industry on poultry farms, it will be more accessible and less expensive to farmers near the facility. Ideally, the use of specific types of biochar on certain soils will also become a common BMP in the region, making farmers more likely to enter nutrient trading schemes. If this is paired with an increase in crop production and a decrease in chemical fertilizer use, it is less likely to put an economic strain on farmers. Many states and regions have already developed both carbon and nutrient trading policies; poultry litter char has potential to be a tradable tool used in both.

References

Carson, E. (2015). Record RGGI Carbon Auction Price Reflects Increasing Market Confidence [online] ENERKNOL Research policy brief. Available at: enerknol.com/record-rggi-carbon-auction-price-reflects-increasing-market-confidence/ [Accessed 4 June 2016].

Chesapeake Bay Program (2017). About the Bay [online]. Available at: www.chesapeakebay.net/about [Accessed 9 July 2017].

Cowie, A., Woolf, D., Gaunt, J., Brandao, M., de la Rosa, R., and Cowie, A. (2015). Biochar, carbon accounting, and climate change. In J. Lehmann and S. Joseph, eds., *Biochar for Environmental Management: Science, Technology And Implementation*, first edition. New York: Routledge, pp. 763–795.

DeBoe, G. and Stephenson, K. (2016). Transactions costs of expanding nutrient trading to agricultural working lands: A Virginia case study. *Ecological Economics*, 130, pp. 176–185.

EnergyWorks (2017). Biopower [online]. Available at: www.energyworks.com/index.php?page=biopower [Accessed 30 June 2017].

EPA (2010). Chesapeake Bay TMDL [online]. Available at: www.epa.gov/chesapeake-bay-tmdl/chesapeake-bay-tmdl-document [Accessed 20 June 2016].

EPA (2015). EPA Global Emissions by Sector [online]. Available at: www.epa.gov/climatechange/ghgemissions/sources/agriculture.html [Accessed 4 June 4 2015].

EPA (2016). National Emissions by Sector [online]. Available at: www.epa.gov/climatechange/ghgemissions/sources/agriculture.html [Accessed 20 July 2016].

EPA (2016). Nitrous Oxide [online]. Available at: www.epa.gov/climatechange/ghgemissions/gases/n2o.html [Accessed 4 June 2016].

EPA (2016). Methane [online]. Available at: www.epa.gov/climatechange/ghgemissions/gases/ch4.html [Accessed 16 December 2016].

Executive Order 13508 (2009). Chesapeake Bay Protection and Restoration [online] Chesapeake Bay Program. Available at: executiveorder.chesapeakebay.net/page/About-the-Executive-Order.aspx [Accessed 20 July 2016].

Federal Register (2015). Carbon pollution emission guidelines for existing stationary sources. *Electric Utility Generating Units*, 80(205), pp 64661–65120, RIN 2066 AR33.

Fisher-Vanden, K. and Olmstead, S. (2013). Moving pollution trading from air to water: potential, problems, and prognosis. *The Journal of Economic Perspectives*, 27(1), pp. 147–171.

Hamilton, D., Cantrell, K., Chastain, J., Ludwig, A., Meinen, R., Ogejo, J., and Porter, J. (2016). Manure treatment technologies: recommendations from the Manure Treatment Technologies Expert Panel to the Chesapeake Bay Program's Water Quality Goal Implementation Team to define manure treatment technologies as best management practices [online]. Chesapeake Bay Program. Accessed at: www.chesapeakebay.net/publications/title/manure_treatment_technologies_bmp_expert_panel_recommendations [Accessed 5 May 2016].

Hoagland, R. A. and Mueller, J. (2011). Legal initiatives driving clean up of Chesapeake Bay. *Maryland Bar Journal*, 44, pp. 13–17.

Jeffery, S., Abalos, D., Spokas, K., and Verheijen, F. (2015). Biochar effects on crop yields. In J. Lehmann and S. Joseph, eds., *Biochar for Environmental Management: Science, Technology And Implementation*, first edition. New York: Routledge, pp. 301–326.

Rabe, B. (2016). Racing to the top, the bottom, or the middle of the pack? The evolving state government role in environmental protection. In N. J. Vig and M. E. Kraft, eds., *Environmental Policy: New Directions for the Twenty-First Century*. Washington DC: CQ/Sage, pp. 33–57.

RGGI (2016). RGGI Offsets [online]. Available at: www.rggi.org/docs/RGGI_Offsets_in_Brief.pdf [Accessed 9 January 2016].

Ribaudo, M. O. and Gottlieb, J. (2011). Point-nonpoint trading – can it work? *JAWRA Journal of the American Water Resources Association*, 47(1), pp. 5–14.

Ribaudo, M. and Savage, J. (2014). Controlling non-additional credits from nutrient management in water quality trading programs through eligibility baseline stringency. *Ecological Economics*, 105, pp. 233–239.

Roberts, K. G., Gloy, B. A., Joseph, S., Scott, N. R., and Lehmann, J. (2009). Life cycle assessment of biochar systems: Estimating the energetic, economic, and climate change potential. *Environmental Science and Technology*, 44(2), pp. 827–833.

Shackley, S., Clare, A., Joseph, S., McCarl, B. A., and Schmidt, H.-P. (2015). Economic evaluation of biochar systems: current evidence and challenges. In J. Lehmann and S. Joseph, eds., *Biochar for Environmental Management: Science, Technology and Implementation*, second edition. New York: Routledge, pp. 813–853.

Shortle, J. S., Ribaudo, M., Horan, R. D., and Blandford, D. (2012). Reforming agricultural nonpoint pollution policy in an increasingly budget-constrained environment. *Environmental Science and Technology*, 46(3), pp. 1316–1325.

Upton, J. (2014). Can we save Chesapeake Bay from chicken crap? [online] Grist. Available at: grist.org/food/can-we-save-chesapeake-bay-from-being-ruined-by-chicken-crap/ [Accessed 25 June 2015].

USDA (2017). Chesapeake Bay Watershed [online]. Available at: www.nrcs.usda.gov/wps/portal/nrcs/detailfull/national/programs/initiatives/?cid=stelprdb1047323 [Accessed 3 March 2017].

Van Houtven, G., Loomis, R., Baker, J., Beach, R., and Casey, S. (2012). Nutrient credit trading for the Chesapeake Bay: An economic study [online]. Chesapeake Bay Commission. Available at: www.chesbay.us/nutrienttrading.htm [Accessed 5 July 2016].

Weibust, I. (2009). *Green Leviathan: The Case For a Federal Role in Environmental Policy*, first edition. New York: Routledge.

Woolf, D., Amonette, J. E., Street-Perrott, F. A., Lehmann, J., and Joseph, S. (2010). Sustainable biochar to mitigate global climate change. Nature Communication [online]. Available at: www.nature.com/ncomms/journal/v1/n5/suppinfo/ncomms1053_S1.html [Accessed 5 July 2016].

2

WATER MANAGEMENT IN THE UPPER KLAMATH BASIN

Collaboration and polarization

Patricia Snyder

Drought pushed the Klamath Basin over the edge and into the national spotlight in 2001. Like many basins throughout the semi-arid and arid American West, generally the Klamath has over-allocated water resources and has, in many ways, been defined by the challenges of managing this scarce resource (Davis, 2001). The water resource landscape in the Klamath and throughout the West represents a labyrinth of complex water law, conflicting demands, and cultural clashes. In 2001, this rural, low-population area was forced center-stage and identified as an example of the water management policy issues faced throughout the American West.

These three big challenges of water law, conflicting demands, and cultural clashes are all interrelated and complicated by the occurrence of drought. In arid and semi-arid systems, drought is a regular occurrence, but global climate models project drought in these regions is likely to occur with greater frequency and greater intensity in coming years (Dettinger, Udall and Georgakakos, 2015). Western water law has its roots in the settling of the West via mining camps and is based on the doctrine of prior appropriation (Davis, 2001). Essentially, this structure ensures that water right holders who received their water right first receive their water before, and at times of great water scarcity in lieu of, junior water right holders (Davis, 2001). This translates into a permitting system in which water right holders are given a priority date for their specific water right, which becomes incredibly important as that date determines who receives full, partial, or no allotments based on water availability in a given year (Beck et al., 1991). Water rights have their origins in the idea of water security; if the investment of building large infrastructure in the arid and semi-arid West was to be made, settlers wanted to ensure that they would have water when needed (Tarlock, 2002). This challenge of complex water law plays into the other challenges already mentioned: the usufructuary nature of water rights makes it an unusual kind of property right, one that essentially gives the holder the right to use the water (Matthews, 2004). This difference between a "true" property right and a water right plays into both conflicting demands and cultural clashes.

The Federal Reclamation movement and building of large water infrastructure by the Bureau of Reclamation (BOR) spurred both agricultural productivity and population expansion (Doremus and Tarlock, 2008). The Klamath Project was built in 1902 and was one of the first under the Federal Reclamation movement (Doremus and Tarlock, 2008). The West was not an empty landscape waiting to be farmed by judicious hands, however; many different tribal communities had a rich and lengthy history already tied to the land. The history of the settlement of the West and its impact on native populations is far too lengthy a discussion to be included here. However, it is important to note that this movement to settle into already existing tribal communities is the foundation of many of the cultural clashes we see today, and has encouraged growth that has yet to abate.

The Klamath encapsulates the conflicting water demands seen throughout the West, including recreational users, irrigators, fish (including commercial and non-commercial interests, and water for healthy habitats and in-stream flow requirements for species listed under the Endangered Species Act), hydropower, tribal communities, and municipalities (Doremus and Tarlock, 2003). The Klamath is a complicated basin because of the arid and drought-prone Upper Klamath; it is also home to the bulk of the farming and ranching in the region, while commercial fisheries reside downstream, meaning whatever is done in the Upper Klamath is felt in the Lower (Powers et al., 2005).

This chapter focuses on the three challenges discussed above associated with water management policy in the Klamath basin, examining them through the lens of specific laws and policies, focusing on the Upper Klamath Basin (Figure 2.1), which straddles the border in south-central Oregon and northern California. The relevant laws and policies encompass various scales. At the federal level, these include the Endangered Species Act (ESA) and the 5th Amendment of the Constitution of the United States. On a regional level, the largest focus will be on the Klamath Basin Restoration Agreement (KBRA) and the Upper Klamath Basin Comprehensive Agreement (UKBCA), two agreements negotiated between stakeholders with historically conflicting demands.

Drought, fish, and a water crisis

In 1988, two species of fish that call the Klamath Basin home were listed under the ESA, requiring more water to stay in the river as in-stream flow (National Research Council, 2004). Things reached a boiling point in 2001 with the occurrence of an extreme drought (Boehlert and Jaeger, 2010). The National Drought Mitigation Center (2017) characterizes an extreme drought by significant agricultural losses and water shortages or restrictions. At the height of the drought, over 82% of the entire Klamath River Basin was categorized as experiencing extreme drought. Drought is a difficult natural hazard to manage for many reasons but mainly because it is multi-modal in nature (Wilhite, Sivakumar and Pulwarty, 2014). In a basin

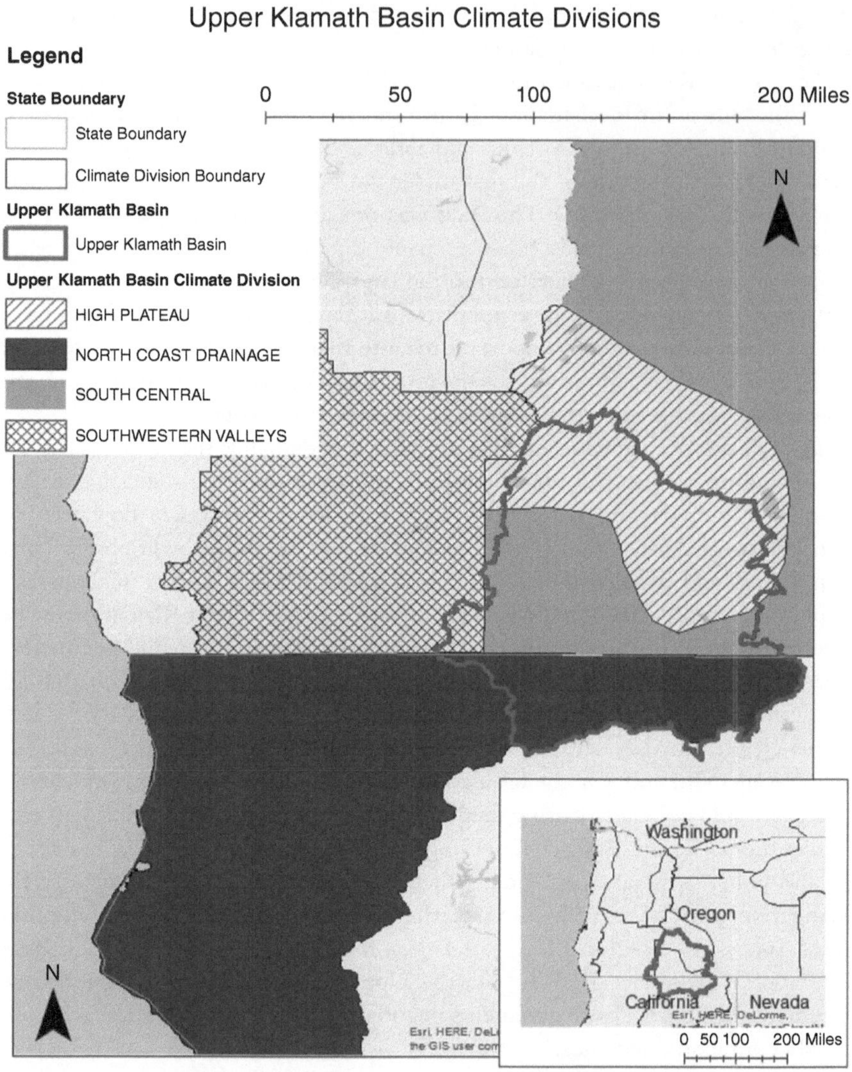

FIGURE 2.1 Upper Klamath Basin climate divisions (Keffer, 2017, created for the author)

where water resources have been historically over-allocated, a drought can drastically complicate water management.

In 2001, for the first time, irrigation allotments were curtailed because of in-stream flow requirements for endangered species: the short-nose and Lost River suckers listed in 1988, as well as the threatened Coho salmon listed in 1997 (Doremus and Tarlock, 2003). The ESA affects water resource allocations in a few ways. Under §7(a)(2) of the Act, consultation between federal agencies is required, meaning that whenever a federal agency and an endangered species are involved in a project, the agency must consult with the United States Fish and Wildlife Service (USFWS) (Parobek, 2003). Prohibited under §9 of the Act, except in situations where incidental take permits are issued (Parobek, 2003), is the taking of an endangered species ("take" is defined in §2(a)(19) as both the harassment and killing/collecting/otherwise removing of a listed species). In the case of aquatic species, the amount of water required to remain in rivers and lakes is determined by federal agencies, which must consult with each other. There is also potential for accidental take when fish end up in irrigation canals (Parobek, 2003).

In spring of 2001, the USFWS and National Marine Fisheries Service (NMFS) finalized their biological opinions regarding the Bureau of Reclamation's (BOR, the federal agency responsible for building most of the large water infrastructure throughout the West) annual operating plan (USFWS, 2001; NMFS, 2001). USFWS recommended that water levels in Upper Klamath Lake be kept higher than usual to prevent die-off of threatened and endangered species of fish (USFWS, 2001). Meanwhile, NMFS recommended that more water be released from Upper Klamath Lake into the Klamath River to maintain higher water levels and lower temperatures for threatened Coho salmon (NMFS, 2001.) Following an extremely dry winter, the snowmelt-dominated Klamath Basin was set to experience extreme drought (Doremus and Tarlock, 2003) during the most water-intensive time of the year: irrigation season. There was not enough water in the basin to satisfy the requirements of both biological opinions, let alone the needs of irrigators (Powers et al., 2005). The end result was a BOR Operations Plan that shut-off all irrigation deliveries from Upper Klamath Lake; in a normal year those would equal between 325,000 and 400,000 acre-feet (Powers et al., 2005).

In an effort to support farmers, then Secretary of the Interior Gale Norton directed 70,000–75,000 acre-feet of water be delivered to irrigators (Powers et al., 2005). This effort was heavily supported by the Bush administration (Doremus and Tarlock, 2008); then Vice President Dick Cheney personally urged scientists in a private meeting to transfer some water to irrigation (Benson, 2002). This amount of water may have helped reduce some of the losses, had it been received earlier in the season. Unfortunately, the timeline of events translated into hundreds of millions of dollars in agricultural losses and affected nearly 100,000 acres, in a devastating economic blow (Powers et al., 2005; Boehlert and Jaeger, 2010).

Parched for a voice

In July of 2001, frustrated irrigators formed "bucket brigades" with a line of people extracting water from Lake Ewauna in buckets, passing them hand-to-hand for over a mile to pour into the BOR's parched A Canal (Owens, 2013). Things escalated until protesters resorted to cutting a fence to force open the valve that feeds A Canal. Law enforcement not only did not intervene but also refused to protect the head gates (Doremus and Tarlock, 2003). Irrigators saw this as a clear battle between farms and fish (Hochberg, 2001). Examples of this cultural clash in the West between ranchers and producers and the federally mandated needs of endangered species and protected areas can be seen both before and after in the Wise Use movement, Sagebrush Rebellion, and occupation of Malheur National Refuge. This conflict stems from a fundamental disagreement on how to manage both wildlife and public lands; water, being necessary for both, makes writing a policy that attempts to take into account all the varied stakeholder demands exceptionally difficult.

The curtailment of irrigation water in 2001 was felt particularly hard by irrigators because it had never happened in the Upper Klamath Basin before. The rest of the nation anxiously watched to see if the Klamath would be able to survive and what that would mean for the state of water management elsewhere in the West. At the height of the crisis, tensions spilled over into violence when three men were arrested for shooting at buildings and signs in Chiloquin, Oregon (Jenkins, 2008). Chiloquin is the heart of the Klamath Tribe reservation, and the men committed this crime while yelling about "sucker lovers," referring to the two endangered species of fish that had been the catalyst for irrigation curtailments (Jenkins, 2008). Following this turbulent summer, Secretary of the Interior Norton had the National Research Council conduct an assessment of both biological opinions, which found that there was not a "substantial scientific basis" for maintaining higher lake levels or for requiring more water to stay in the Klamath River (National Research Council, 2004). These findings have been hotly debated, and both the report itself and the original biological opinions remain sensitive subjects in the Upper Klamath, resulting in the addition of mistrust of science in a culture that already greatly distrusts federal government and its perceived overreach (National Research Council, 2008).

The pendulum shifts

The following year, USFWS and NMFS offered "letters of concurrence" allowing for regular water deliveries, and in April flow below Iron Gate Dam was reduced to 1,350 cubic feet per second (cfs) (Powers et al., 2005). At the height of irrigation season, the flow from Iron Gate Dam fell to 760 cfs (Guillen, 2003). NMFS had recommended a minimum flow of 1,700 cfs in its 2001 Biological Opinion (NMFS, 2001). Debates and lawsuits raged over the appropriate amount of water that should both stay in Upper Klamath Lake and be released downstream to support fisheries further down the Klamath River. In May 2002, both agencies released Biological Opinions criticizing the BOR's ten-year operation plan, purporting that it would likely negatively affect endangered and threatened species (USFWS, 2002; NMFS, 2002). The agencies outlined "reasonable and prudent alternatives" that would compromise between healthy fisheries and the BOR's operations (USFWS, 2002; NMFS, 2002).

The BOR rejected these alternatives and compiled its own one-year plan, claiming it complied with both Biological Opinions (Powers et al., 2005). Despite this claim, in September of 2002, over 30,000 adult Chinook salmon died (not the federally listed Coho but still important for commercial fisheries) (California Department of Game and Fish, 2004). There were some who thought the political pressure applied by the Bush administration to ensure water for irrigators (Doremus and Tarlock, 2008) resulted in the BOR not utilizing the best available science in regards to the health of the fishery (Daley, 2007). The cause of this massive fish kill was epizootic disease encouraged by three factors: 1) an unusually large number of Chinook salmon migrating up the Klamath River from the Pacific Ocean; 2) below average flow and water volume in the Klamath River; and 3) warm water temperatures (California Department of Game and Fish, 2004). It is still debated how much, or if at all, BOR's 2002 Operation Plan contributed to this fish kill, but it certainly did nothing to alleviate tensions in the basin. The Yurok Tribe of the Lower Klamath Basin completed their own analysis of the fish kill and established that had flow below Iron Gate Dam been increased, specifically to consistently >1,000 cfs, the fish kill likely would not have occurred (Belchick, Hillemeier and Pierce, 2004). Fish play an integral and complex role in this basin, contributing to economic, spiritual, and cultural well being. The effect of water policy on these events was the beginning of settlement negotiations between stakeholders.

A way forward?

Negotiated agreements surrounding water had been attempted before in the Klamath Basin, but the water crisis and fish kill of 2001–2002 solidified in many stakeholders' minds the necessity for such an agreement. With the likely increase in occurrences of drought, combined with the legal requirements through the ESA, water management policy required more effective tools. Although the events of 2001 and 2002 are certainly provocative and attention capturing, it is perhaps this realization and the negotiating efforts that followed that are the most extraordinary. In 2005, various stakeholder groups who had historically been unwilling to compromise with each other came to the negotiating table, sat down, and began to talk.

It took another five years for a draft agreement to be finalized. In 2010, representatives of the states of California and Oregon; the Karuk, Klamath, and Yurok Tribes; Del Norte, Siskiyou, and Humboldt counties in California; Klamath County in Oregon; the Klamath Reclamation Project; Upper Klamath irrigators; and seven fish and/or environmental organizations signed a landmark agreement: the Klamath Basin Restoration Agreement (KBRA) (see Figure 2.2) (Klamath Basin Restoration Agreement, 2010).

There are four main components to the KBRA: 1) restoration of riparian areas throughout much of the basin; 2) the removal of four dams: J.C. Boyle, Copco 1 and 2, and Keno; 3) the acquisition of the Mazama Forest on behalf of the Klamath Tribe; and 4) assurances to irrigators to provide additional water security as well as assistance with infrastructure (KBRA, 2010). Two other agreements were formed following the signing of the KBRA: the Klamath Hydroelectric Settlement Agreement (KHSA) and Upper Klamath Basin Comprehensive Agreement (UKBCA) (see Table 2.1).

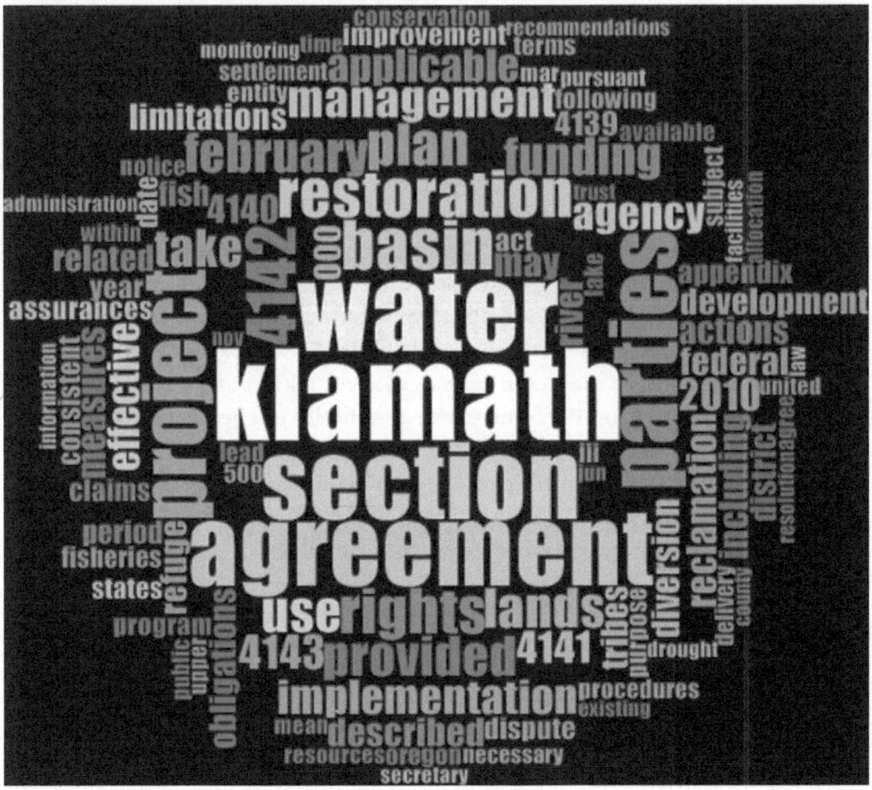

FIGURE 2.2 Word cloud of most used words within the Klamath Basin Restoration Agreement

The signing of this agreement was monumental; it was not, however, universally supported, with some even referring to it as the "Klamath Basin Surrender Act" (Glucklich, 2010). Recognizing that it was impossible to appease everyone, stakeholders who were less willing to compromise were excluded, leaving many feeling that the agreement was forged in shadows and "closed-door meetings" (Floyd, 2016; WaterWatch, 2017). The main points of contention surrounded the removal of dams and the acquisition of the Mazama Forest. This acquisition would be made on behalf of the Klamath Tribes and, because of the compromises that were made on the part of the tribes over water, was viewed by many as trading land for water.

Back and forth

The Upper Basin irrigators generally opposed the KBRA. The main contentions centered on requirements under the ESA, water security, and purchase of land for the Klamath Tribe (Krizo, 2009). The first and second issues are intrinsically tied. Irrigators worried that even if the KBRA was ratified, the ESA would ultimately

TABLE 2.1 Four main Klamath Basin agreements and their attributes

Name	*Year*	*Requires Federal Authorizing Legislation*	*Status*	*Main Points*
Klamath Basin Restoration Agreement	2010	Y	Terminated	• Restoration of riparian area • Removal of four dams • Acquisition of the Mazama Forest on behalf of the Klamath Tribe • Assurances to irrigators regarding water security/ concerns regarding take of listed species
Klamath Hydroelectric Settlement Agreement	2010	Y	Active but negotiations ongoing to separate from KBRA	• Establishes the necessity for studies regarding the feasibility of dam removal
Upper Klamath Basin Comprehensive Agreement	2014	Y	Active but negotiations ongoing to separate from KBRA	• Water use program: increase flow to Upper Klamath Lake by ~30,000 acre-feet with input from irrigators • Riparian program: restore and protect riparian zones • Economic development program: create economic opportunities for the Klamath Tribe • Transition program: meant to ease burden by new arrangements • Establishes Landowner Entity, who oversees Water Use and Riparian programs • Establishes Joint Management Entity, comprised of Klamath Tribes, State of Oregon, and Landowner Entity

continued

TABLE 2.1 *(cont.)*

Name	*Year*	*Requires Federal Authorizing Legislation*	*Status*	*Main Points*
Klamath Power and Facilities Agreement	2016	Y	Active	• Helps to protect on-project irrigators from "take" consequences of listed species due to a) dam removal or b) riparian restoration • Non-federal stakeholder groups commit to source funding to construct fish screens • Non-federal stakeholder groups commit to support legislation for lower power rates for irrigators

limit water availability and many felt if the federal government had not been able to recover the threatened and listed species thus far, a new agreement would not help (Hearden, 2011). Many within the region also felt that tribal termination policies of the mid-twentieth century paid the Klamath for their land and purchasing the Mazama Forest was unnecessary (Kelly, Bliss, and Gosnell, 2013). This sentiment can be felt in this quote from the then chairman of the Klamath Basin Alliance: "There is a bigger question of the great injustice of Tribal sovereignty where the tribes have used the endless checkbook of the federal government, attacking our agricultural community" (Bayona, 2002; Kelly, Bliss, and Gosnell, 2013).

Many irrigators felt that regardless of what the agreement specified, the ESA would still dictate the amount of water available for irrigation (Beaver, 2010). The legal framework surrounding water in the West, which has centered on the idea of water as a property right, also creates conflict. Curtailing water allotments is often seen as an overreach of the federal government (Parobek, 2003). This cultural viewpoint manifests itself legally in the idea of takings and has resulted, in the Klamath, in divisive lawsuits.

The Fifth Amendment of the United States Constitution, more commonly referred to as the Takings Clause, protects property owners from both federal government seizures and the loss of economic viability based on interpretations of the statement "nor shall private property be taken for public use, without just compensation." Case law has created two distinct types of takings evaluations, one involving the physical seizure of property; the second referred to as "regulatory taking" and subject to a balancing test (Echeverria, 2005).

Penn Central v. City of New York (1978) created a three-part balance test: courts must consider 1) the economic impact of the government's action; 2) how much said action has disrupted the plaintiff's investment-backed expectations; and 3) the character of the government's action, generally interpreted in terms of reasonableness (Echeverria, 2005). *Lucas v. South Carolina* (1992) established the exception of a categorical taking, where federal government action has robbed property of all economic value (Westbrook, 2006). The importance of this in terms of water resources and stakeholders within the Upper Klamath Basin can be seen in *Klamath Irrigation District v. United States* (2005).

The Klamath Irrigation District case highlights the differences in demands between various stakeholders within the Upper Klamath Basin: "The Klamath litigation highlights the ongoing cultural war that is waging in the American West. It showcases the battle between the status quo of the irrigation culture and the changes in demand that have occurred in response to the booming populations of western cities" (Dunlap, 2013, p. 114). Following the drought of 2001, irrigators within the basin filed suit against the federal government. The complaint hinged on whether the curtailment of irrigation deliveries was a) a takings and b) a breach of contract under the Klamath River Basin Contract (1957), the agreement between California and Oregon regarding water resources for the basin (Westbrook, 2006). The original takings case was dismissed but upon appeal the case has been revived (Spohr, 2012); a decision in the case is expected by 2018.

In the West, land and water are closely tied both economically and culturally. The Klamath reservation was all but eliminated by post-World War II termination policies; today all that remains is 372 acres shared between the Klamath, Modoc, and Yahooskin tribes (Doremus and Tarlock, 2008). Termination policies were put into place in an attempt to force tribes to assimilate more quickly and be "freed from Federal supervision and control" (Walch, 1983, p. 1185). The Klamath Termination Act of 1954 dissolved the Klamath reservation (Hood, 1972). However, *United States v. Adair* (1983) solidified the water rights of the tribe, even without a reservation, by confirming the water rights of the tribe to protect their hunting and fishing rights on former reservation lands as well as securing enough for agricultural production. The status of the Klamath as a tribe essentially without a reservation is unusual, and the loss of land has been severely felt by them (Doremus and Tarlock, 2008). The precedent set by *United States v. Adair* has resulted in the legal system placing these rights in "time immemorial," meaning they have first priority (Doremus and Tarlock, 2008, p. 72). A 1908 case, *Winters v. United States*, established the Reserved Right Doctrine, which ensures that when the federal government reserves land for a specific purpose, it is also implicitly reserving water rights necessary to fulfill that purpose (Benson, 2002). While this is often referred to in terms of tribal lands, it also provides water rights for other federal reserved lands, like the Lower Klamath National Wildlife Refuge, which, in practice, is generally last in line to receive water.

In 2013, this dynamic shifted through adjudication on the Oregon side of the Upper Klamath Basin. Adjudication is the process by which water rights are officially given a priority date and quantified, which began for the entire state of

Oregon in 1975 (Milner, 2015). Through the Final Order of Determination in 2013, the Klamath Tribes had the largest right in the basin, with the oldest priority date and, most importantly, it was now enforceable (Cosens and Chaffin, 2016). This event acted as a catalyst in the Upper Klamath Basin. Upper Basin irrigators, who had been unwilling to negotiate previously, were now much more inclined to do so, as they realized it was possible they would end up with far less water than they were accustomed to, or potentially none at all. The Upper Klamath Basin Comprehensive Agreement (UKBCA) was the result: an agreement between stakeholders throughout the upper portion of the Klamath Basin. The four main elements of this agreement include: 1) a water use program, which essentially increases inflows to Upper Klamath Lake while also providing water for irrigators; 2) a riparian program, which both protects and restores riparian zones; 3) an economic program, which exists to increase economic opportunities for the Klamath Tribe; and 4) a transition program, which essentially creates a blueprint for how these programs will be established over time (UKBCA, 2014).

The energy-water-food nexus

The energy-water-food nexus in the Upper Klamath cannot be ignored. With the advent of large water infrastructure and the Federal Reclamation movement came low power rates for farmers within the Klamath (Powers et al., 2005). The BOR negotiated with Copco (a predecessor of the current energy company, PacifiCorp) (Jaeger, 2004) an agreement that essentially resulted in the building of Link River Dam on PacifiCorp's dime, which they would then turn over to BOR for ownership while committing to providing low energy rates for farmers within the basin for 50 years (Jaeger, 2004). In exchange, PacifiCorp received the rights to generate hydropower on the Klamath River; these low energy rates translate into about one-tenth of what other irrigators pay (Jaeger, 2004). In 2006, PacifiCorp did not renew the contract that had been in place nearly 100 years and began the process to raise energy costs, a ten-fold increase implemented annually, through negotiations with the Public Utilities Commissions in Oregon and California (Souza, 2006; Doremus and Tarlock, 2008).

The rate increases were essentially caused by aging infrastructure, the Federal Power Act, the Clean Water Act, and a disinclination of state and federal government to put money into large water infrastructure projects as had been done in the past. In 2006, the Klamath Hydropower Project was up for re-licensing through the Federal Energy Regulatory Commission (FERC); however, PacifiCorp realized re-licensing would require added costs to construct fish passage as required under the Federal Power Act and changes to reservoirs that would allow for 401 Water Quality Certification under the Clean Water Act (US DOI, US DOC, and NMFS, 2013). These added costs would have been passed on to the customer and may not have been fully recouped by PacifiCorp for some time.

The four dams slated for removal have a combined maximum storage capacity of just over 96,500 acre-feet and have a maximum power capacity of 163 MW

TABLE 2.2 Benefits to dam removal as outlined in the KHSA

Benefit	*Total Amount*	*Percentage Increase*
Commercial fishing for Chinook salmon benefit	$134.5 million	43.0%
Sport and tribal harvests	$1.75 million	6.2%
Increase in water to irrigators during drought years	$29.89 million	1.9%
Refuge Recreation	$4.3 million	46.7%

Source: US DOI, US DOC, and NMFS, 2013

(US DOI, US DOC, and NMFS, 2013). The dam with the most storage capacity is Iron Gate, with nearly 54,000 acre-feet of maximum storage available, found at river mile 190 (US DOI, US DOC, and NMFS, 2013). The dam with the highest maximum power capacity is JC Boyle, with 98 MW, found at river mile 224 (US DOI, US DOC, and NMFS, 2013). The KHSA was signed in conjunction with the KBRA to explore the ecological and socio-economic impacts of dam removal. The associated report outlines the benefits of dam removal and lists a number of specific economic benefits (Table 2.2). It should be noted that sport and tribal harvesting summarizes benefits for salmon populations alone and all figures are in 2012 dollars, excepting cost estimates, which are in 2020 dollars. The numbers that address benefits are directly tied to the KBRA and its associated plans, commitments, and programs and do not necessarily reflect benefits of dam removal without these (US DOI, US DOC, and NMFS, 2013).

The low cost-estimate for full removal of the four dams is $238 million, while the high cost-estimate is $493 million, which exceeds the state cost cap of $450 million (US DOI, US DOC, and NMFS, 2013). The analysis pinpoints the most likely amount to be just over $291.5 million; however, there are mechanisms within the KHSA that would address a cost above the state cost cap (Klamath Hydroelectric Settlement Agreement, 2010).

KBRA dissolves

In 2015, after five years of effort, the KBRA sunset due to a lack of congressional action, which re-polarized many of the stakeholder groups. It also put new energy into those groups who had opposed any negotiated agreement from the beginning. Dam removal, in particular, was one of the most controversial items within the agreement but, ironically, it may be one of the very few that survives it.

In 2016, then Secretary of the Interior Sally Jewell announced the process of removing the four dams discussed in the KBRA would begin (Department of the Interior, 2010). The decision culminated with the signing of a new document, the Klamath Power and Facilities Agreement (KPFA), which supports the removal of all four dams and is considered "a step in the direction" to address irrigator concerns (Klamath Power and Facilities Agreement, 2016) but lacks the concrete compromises made to benefit irrigators found in the KBRA. The dam removal process, with a projected completion by 2020, will go through the FERC re-licensing process. It is unclear how the Trump administration appointees to the commission, whose approval is necessary to move forward, will view dam removal.

Conclusion

The events discussed within this chapter have acted as catalysts for policy change in the Upper Klamath Basin (see Figure 2.3). The first catalyst was the drought of 2001, which pushed strategies (ie, the curtailment of irrigation deliveries) towards prioritizing endangered and threatened fish. The following year, political pressure was able to swing the pendulum the other way and allow for more water to be delivered to irrigators, which may have influenced massive fish kills in 2002. The 2006 energy rate hike was another catalyst event that transformed water management and the relationships among stakeholders. This event pushed many irrigators in the basin further away from a compromise due to feeling that their needs were not being met and their lifestyle was under direct attack via the increasing difficulty of making a profit on agriculture and ranching. The negotiation and signing of the KBRA was both a result of the first two catalyst events, bringing some stakeholders together, and a catalyst itself, pushing those not at the negotiation table farther away. The 2013 Final Order of Determination on the Oregon side of the Upper Klamath Basin represented a power shift, whereby the Klamath Tribe now had the oldest and largest enforceable water right. This event encouraged the Upper Basin irrigators to participate in negotiations and resulted in the UKBCA, also representing a power

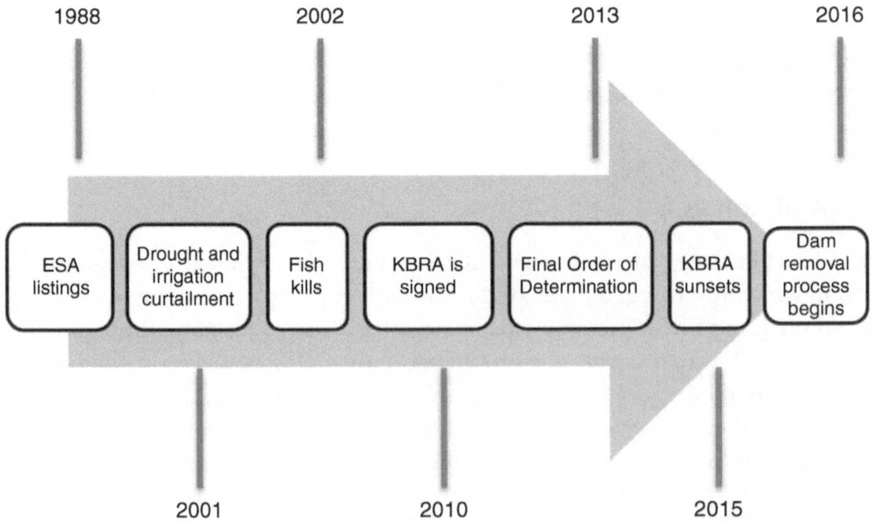

FIGURE 2.3 Timeline of catalyst events affecting water policy in the Upper Klamath Basin

Source: Patricia Snyder

shift within the basin. The termination of the KBRA is perhaps the most significant catalyst event in the basin; it has re-polarized those stakeholders who were working together and re-energized those opposed to a negotiated agreement. The final catalyst event discussed is the potential removal of dams, and dam removal remains a polarizing subject. These catalyst events show the extent to which events on the socio-political landscape affect the implementation of water policy and management as much as the availability of water itself and climatic and geographic realities.

The efforts made on local and regional levels have impacted the basin and created relationships among stakeholders who have historically had conflicting ideas on how water should be managed. The next few years, however, are likely to be difficult ones for the Klamath; the re-polarization of stakeholder groups and discouragement felt after the dissolution of the KBRA and the ongoing dam removal process are likely to have ripple effects. The dissolution of the KBRA also shows that efforts made on local and regional levels are not enough; although local collaboration is paramount to successful water management in any basin in the West, more capacity is necessary. The compromises that were made in the KBRA needed not only federal ratification but also federal funding to come to fruition. The takeaway for water management in the West is that broad-scale solutions that are necessary for many basins like the Klamath must be accompanied by broad-scale support and strong political will. Federal policy and legislation has effects on local and regional resource management; these effects must be recognized and mitigated through policy and legislation, as well as funding, or the efforts of local and regional stakeholders will be wasted.

References

Bayona, L. (2002). The case against restoring the reservation. *Oregon Herald and News*, p. A1.

Beaver, T. (2010). After drought, irrigators still split on KBRA. *Oregon Herald and News*, p. A3.

Beck, R., Anderson, O., Kuntz, E., and Simmons, P. (1991). Elements of prior appropriation. In *Waters and Water Rights*, ed. R. Beck, pp. 12–68. New York: Lexis Publishing.

Belchick, M., Hillemeier, D., and Pierce, R. (2004). The Klamath River fish kill of 2002; Analysis of contributing factors. Yurok Tribal Fisheries Program. Available from: www.yuroktribe.org/departments/fisheries/documents/FINAL2002FISHKILLREPORTYTFP.pdf [Accessed 14 August 2017].

Benson, R. (2002). Giving suckers (and salmon) an even break: Klamath Basin water and the Endangered Species Act. *Tulane Environmental Law Journal*, 15(2), pp. 197–238.

Boehlert, B. and Jaeger, W. (2010). Past and future water conflicts in the Upper Klamath Basin: An economic appraisal. *Water Resources Research*, 46 (10). Available from: onlinelibrary.wiley.com/doi/10.1029/2009WR007925/pdf [Accessed 14 August 2017].

California Department of Game and Fish (2004). September 2002 Klamath River fish-kill: final analysis of contributing factors and impacts. Available at: waterwatch.org/wp-content/uploads/2011/08/2002CAFinalKlamathFishKillReport.pdf [Accessed 14 August 2017].

Cosens, B. and Chaffin, B. (2016). Adaptive governance of water resources shared with indigenous peoples: the role of law. *Water*, 8(3), p. 97.

Daley, M. (2007). Dems: Investigate Cheney for role in salmon die-off. *The Seattle Times*. Available from: www.seattletimes.com/seattle-news/dems-investigate-cheney-for-role-in-salmon-die-off/ [Accessed 14 August 2017].

Davis, S. (2001). The politics of water scarcity in the Western states. *Social Science Journal*, 38, pp. 527–542.

Department of the Interior (2010). Secretary Salazar, Governors Kulongoski and Schwarzenegger announce agreement on Klamath River Basin restoration. Available from: www.doi.gov/news/photos/Klamath-River-Basin-Restoration-Agreement [Accessed 14 August 2017].

Dettinger, M., Udall, B., and Georgakakos, A. (2015). Western water and climate change. *Ecological Applications*, 25(8), pp. 2069–2093.

Doremus, H. and Tarlock, A. (2003). Fish, farms, and the clash of cultures in the Klamath Basin. *Ecology Law Quarterly*, 30, pp. 279–350.

Doremus, H. and Tarlock, A. (2008). *Water War in the Klamath Basin: Macho Law, Combat Biology, and Dirty Politics*. Washington, DC: Island Press.

Dunlap, J. (2013). Does first in time really mean first in right? Exploring water rights in the context of Klamath Irrigation District v. United States. *Journal of Environmental Law and Litigation*, 28(2), pp. 109–150.

Echeverria, J. (2005). Making sense of Penn Central. *Journal of Environmental Law and Policy*, 23(2), pp. 171–210.

Floyd, S. (2016). Dam removal across the West a growing concern. *Herald and News* [online]. Available from: www.heraldandnews.com/news/dam-removal-across-the-west-a-growing-concern/article_358fe4d6-317c-592a-88f0-f66099315d54.html [Accessed 14 August 2017].

Glucklich, E. (2010). Tea Party protesters speak out. *Herald and News* [online]. Available from: www.heraldandnews.com/top_story/tea-party-protesters-speak-out/article_a32d9808-491e-11df-8f00-001cc4c03286.html [Accessed 14 August 2017].

Guillen, G. (2003). *Klamath River Fish Die-Off September 2002: Report on Estimate Mortality*. Arcata, CA: USFWS.

Hearden, T. (2011). Klamath divisions persist. Capital Press [online]. Available from: www.capitalpress.com/content/TH-klamath-hearing-w-photos-video-infoboxes-102111 [Accessed 14 August 2017].

Hochberg, L. (2001). Fish vs. farmers. Public Broadcasting System [online]. Available at: www.pbs.org/newshour/bb/environment-july-dec01-fish_8-20/ [Accessed 14 August 2017].

Hood, S. (1972). Termination of the Klamath Indian Tribe of Oregon. *Ethnohistory*, 19(4), pp. 379–392.

Jaeger, W. (2004). Energy pricing and irrigated agriculture in the Klamath Basin. Oregon State University Extension Catalog [online] 8846-E. Available at: catalog.extension.oregonstate.edu/em8846 [Accessed 14 August 2017].

Jenkins, M. (2008). Peace on the Klamath. High Country News. Available at: www.hcn.org/issues/373/17763 [Accessed 14 August 2017].

Keffer, M. (2017). Upper Klamath Basin climate divisions. Created for the author.

Kelly, E., Bliss, J., and Gosnell, H. (2013). The Mazama returns: the politics and possibilities of tribal land reacquisition. *Journal of Political Ecology*, 20, pp. 429–433.

Klamath Basin Restoration Agreement (2010). Available at: klamathrestoration.gov/sites/klamathrestoration.gov/files/Klamath-Agreements/Klamath-Basin-Restoration-Agreement-2-18-10signed.pdf [Accessed 14 August 2017].

Klamath Hydroelectric Settlement Agreement (2010). Available at: 216.119.96.156/Klamath/Klamath%20Hydroelectric%20Settlement%20Agreement%202-18-10signed.pdf [Accessed 14 August 2017].

Klamath Power and Facilities Agreement (2016). Available at: www.doi.gov/sites/doi.gov/files/uploads/REVISED%204-6-2016%20Yurok%20DRAFT%202016%20Klamath%20Power%20%26%20Facilities%20Agrmt%20%20CLEAN.pdf [Accessed 14 August 2017].

Krizo, J. (2009). Senator Wyden listens to water concerns. Capital Press. Available at: klamathbucketbrigade.org/CapitalPress_Wydenlistenstowaterlawconcerns030609.htm [Accessed 14 August 2017].

Matthews, O. (2004). Fundamental questions about water rights and market reallocation. *Water Resources Research*. Available at: onlinelibrary.wiley.com/doi/10.1029/2003WR002836/epdf [Accessed 26 October 2017].

Milner, M. (2015). Water law meets participatory democracy: a Klamath Basin example. *Journal of Environmental Law and Litigation*, 30(1), pp. 87–142.

National Drought Mitigation Center, US Department of Agriculture, US Department of Commerce, and National Oceanic Atmospheric Administration (2017). US Drought Monitor. Available at: droughtmonitor.unl.edu/MapsAndData/DataTables.aspx [Accessed 14 August 2017].

National Marine Fisheries Service (2001). *Biological Opinion: Klamath Project Operations.* Long Beach, CA: National Marine Fisheries Service Southwest Region.

National Marine Fisheries Service (2002). Biological opinion: Klamath project operations. Available at: www.westcoast.fisheries.noaa.gov/publications/Klamath/nmfs_2002_klamath_project_bo.pdf [Accessed 14 August 2017].

National Research Council (2004). *Endangered and Threatened Fishes in the Klamath River Basin: Causes of Decline and Strategies for Recovery.* Washington, DC: National Academies Press.

National Research Council. (2008). *Hydrology, Ecology, and Fishes of the Klamath River Basin.* Washington, DC: National Academies Press.

Owens, H. (2013). 2001 Bucket Brigade. *Herald and News* [online]. Available at: www.heraldandnews.com/bucket-brigade/image_21e38828-de96-11e2-8662-001a4bcf887a.html [Accessed 14 August 2017].

Parobek, C. (2003). Of farmer's takes and fishes' takings: Fifth Amendment compensation claims when the Endangered Species Act and Western water rights collide. *The Harvard Environmental Law Review*, 27(1), pp. 177–225.

Powers, K., Baldwin, P., Buck, E., and Cody, B. (2005). Klamath River Basin issues and activities: an overview. Congressional Research Service [online] RL33098. Available at: nationalaglawcenter.org/wp-content/uploads/assets/crs/RL33098.pdf [Accessed 14 August 2017].

Souza, C. (2006). Huge power rate hike hits Klamath farms. AgAlert. Available at: www.agalert.com/story/?id=575 [Accessed 14 August 2017].

Spohr, D. (2012). (When) does a contract claim trump a takings claim? Lessons from the water wars. *Washington Journal of Environmental Law and Policy*, 2(1), pp. 125–168.

Tarlock, D. (2002.) The future of prior appropriations in the West. *Natural Resources Journal*, 41, pp. 769–793.

Upper Klamath Basin Comprehensive Agreement (2014). Available at: klamathtribes.org/wp-content/uploads/2014/08/2014-4-18-UPPER-KLAMATH-BASIN-COMPREHENSIVE-AGREEMENT.pdf [Accessed 14 August 2017].

US Department of Interior, US Department of Commerce, and National Marine Fisheries (2013). Klamath Dam removal overview, report for the Secretary of the Interior: an assessment of science and technical information. Available at: klamathrestoration.gov/sites/klamathrestoration.gov/files/2013%20Updates/Final%20SDOR%20/0.Final%20Accessible%20SDOR%2011.8.2012.pdf [Accessed 14 August 2017].

US Fish and Wildlife Service (2001). Biological/conference opinion regarding the effects of operation of the BOR's Klamath Project on the endangered Lost River sucker, endangered shortnose sucker, and threatened bald eagle, and proposed critical habitat for the Lost River/shortnose suckers. Klamath Falls, OR: US Fish and Wildlife Service.

US Fish and Wildlife Service (2002). Biological/conference opinion regarding the effects of operation of the BOR's Klamath Project on the endangered Lost River and shortnose suckers. Sacramento, CA: USFWS Pacific Southwest Region.

US Fish and Wildlife Service (2004). The Klamath River Basin. Available at: www.fws.gov/yreka/Maps/KlamathRvBasinDEM.jpg [Accessed 14 August 2017].

US Geological Survey (2014). Evaluation of alternative groundwater-management strategies for the BOR Klamath Project, Oregon and California. USGS [online]. Available at: pubs.usgs.gov/sir/2014/5054/pdf/sir2014-5054.pdf [Accessed 14 August 2017].

Walch, M. (1983). Terminating the Indian Termination Policy. *Stanford Law Review*, 35(6), pp. 1181-1215.

WaterWatch (2017). The Klamath Settlement Agreements. Available from: waterwatch.org/updates/7 [Accessed 14 August 2017].

Westbrook, E. (2006). Recognizing the limits of water rights: rejecting takings claims in Klamath Irrigation District v. United States. *Environmental Law Reporter* [online] 36, pp. 1026–1036. Available at: www.eli.org/sites/default/files/docs/westbrook_2006.pdf [Accessed 14 August 2017].

Wilhite, D., Sivakumar, M., and Pulwarty, R. (2014). Managing drought risk in a changing climate. *Weather and Climate Extremes*, 3, pp. 4–13.

3

TOWARD A WHOLE-OF-GOVERNMENT AND WHOLE-OF-COMMUNITY APPROACH FOR REGIONAL ADAPTATION TO SEA LEVEL RISE

Lessons learned from the Hampton Roads Intergovernmental Pilot Project

Juita-Elena (Wie) Yusuf, Michelle Covi, Carol Considine, Burton St. John III, Meagan M. Jordan, and J. Gail Nicula

Sea level rise (SLR) and climate change, like many other environmental issues, are wicked problems that are constantly evolving and changing (Jentoft and Chuenpagdee, 2009; Kreuter et al., 2004; Lach, Rayner, and Ingram, 2005; Lazarus, 2008; Levin et al., 2012; Moser et al., 2012). Planning for SLR in coastal communities is a challenging problem because of the political polarization associated with climate change (Grundmann, 2007; Hamilton, 2011; McCright and Dunlap, 2011; Yusuf et al., 2014), because of the lack of effectiveness of incremental solutions (Moser et al., 2012), and because it takes place in an environment of "sector failure" (Bryson, Crosby, and Stone, 2006, p. 46) where one sector alone – whether government, business, or society – cannot effectively pursue adaptation, yet collaboration is difficult. Planned adaptation to SLR requires collaborative, multi-sectoral, and multi-jurisdictional approaches (Adger et al., 2005).

SLR is not bound by geographic boundaries, but governments at all levels – local, state, and federal – generally limit their policies and adaptation responses, such as vulnerability assessments, coastal management and planning, and adjustment to infrastructure, to their jurisdictional authority. However, effective adaptation to SLR requires considering functional boundaries of ecosystems rather than political boundaries. Pursuing planned adaptation using a whole-of-government approach requires coordination across different levels of government (intergovernmental) and departments and agencies within the same levels of government (intragovernmental), as risks are not constrained by legal, geopolitical, or sector boundaries.

Planned adaptation to SLR, such as through zoning, land use, or infrastructure policies, requires that multiple sectors – government, business, non-profit, and civil society – be involved. This whole-of-community approach emphasizes the value and importance of strengthening relationships and communication among the wide range of stakeholders, including local, regional, state, and federal governments; non-governmental, faith-based, and non-profit organizations; the private sector; educational, health care, and other institutional stakeholders; and individuals, families, and communities (Federal Emergency Management Agency, 2011).

The goal of planned adaptation is to enhance socio-ecological resilience, defined as "the capacity of linked social-ecological systems to absorb recurrent disturbances such as hurricanes or floods so as to retain essential structures, processes, and feedbacks" (Adger et al., 2005, p. 1036). Building resilience to natural disasters in an era of climate change requires developing cooperative systems across society. Focusing on socio-ecological resilience offers a way to pursue adaptation in a whole-of-government and whole-of-community approach that avoids multi-jurisdictional and multi-sectoral conflict. Socio-ecological resilience requires understanding adaptation at broader scales (Adger et al., 2005) and involves multi-level and multi-sectoral networks to support the political, legal, and economic frameworks of effective governance (Dietz, Ostrom, and Stern, 2003; Folke et al., 2005). It calls for institutional arrangements that allow coordinated, collaborative planning that spans multiple watersheds, political jurisdictions, and boundaries.

It is not unusual for policymaking and governance structures and ecological systems to involve a mismatch of scales, such that the scale of environmental variation and the scale of policy and management responsibility result in disruption of one or more functions of the socio-ecological system (Cumming, Cumming, and Redman, 2006). Resolving the scale mismatch requires institutional reorganization by creating governance structures that allow scalar flexibility (Borgström et al., 2006). This chapter discusses the case study of one institutional arrangement – the Hampton Roads Intergovernmental Sea Level Rise and Resilience Pilot Planning Project (known as the "Pilot Project"). The Pilot Project sought to overcome scale mismatch through a two-year, whole-of-government and whole-of-community effort involving multiple municipal governments, state and federal agencies, academic institutions, non-profit organizations, businesses, civic, and neighborhood organizations, and individual residents. This case offers lessons for environmental policy in pursuit of sustainability by highlighting challenges in pursuing regional level (ie, sub-state, supra-local) policies that overcome scale mismatches through a multi-sectoral, multi-governmental framework. The case also demonstrates how the gap between policy intentions and policy output is widened by goal ambiguity, role confusion, and resource inadequacies.

Case study background

The Hampton Roads Intergovernmental Sea Level Rise and Resilience Pilot Planning Project (the Pilot Project) is located in the Hampton Roads region of

FIGURE 3.1 Map of Hampton Roads, Virginia
Source: Yusuf et al., this chapter

southeastern Virginia (USA) at the confluence of the Chesapeake Bay and Atlantic Ocean (see Figure 3.1). The region is comprised of 17 municipal governments and includes 26 federal facilities. The region also houses several state facilities and infrastructure.

The Hampton Roads area is vulnerable to extensive asset damage due to accelerated SLR (Atkinson, Ezer, and Smith, 2013; Ezer and Atkinson, 2014; Governor's Commission on Climate Change, 2008; Hampton Roads Planning District Commission, 2010; 2012; Kleinosky, Yarnal, and Fisher, 2007; Li, Lin, and Burks-Copes, 2012; McFarlane, 2013). Potential threats include risks to transportation and other public infrastructure, ports and logistics, military operations, tourism, wetlands, and coastal ecosystems (Hampton Roads Planning District Commission, 2010; Pyke et al., 2008; Wu, Najjar, and Siewert, 2009). Like many other vulnerable areas, the region faces an implementation deficit (Yusuf and St. John III, 2017) with possible solutions having not been translated into implementation.

Regionalism and regional coordination have consistently been a challenge. Each locality typically plans only within its jurisdictional boundaries, and there are few higher level, supra-local mechanisms that enable or support regional collaboration on infrastructure improvements. Political and historical influences have fostered fragmentation, as municipalities value their identity, independence, and autonomy. Hampton Roads has several regional entities and authorities, including the Hampton Roads Sanitation District, Southeastern Public Service Authority, Hampton Roads Transit, and Hampton Roads Planning District Commission (HRPDC). While some regional agreements have been successful, parties are largely reluctant to surrender their autonomy. Thus, the region faces challenges in achieving regional cooperation and coordination, largely due to the sheer number of jurisdictions involved, and political and legal factors (Babcock, 2009; Bacon, 2015).

Pilot Project overview

The Pilot Project was an effort to "use the knowledge, skills and expertise of all regional stakeholders to create a framework or template for intergovernmental strategic planning that could be used outside the region and to implement that integrated strategy in Hampton Roads, Virginia, creating an effective and efficient method for planning holistically for sea level rise and recurrent flooding" (Hampton Roads Sea Level Rise Preparedness and Resilience Intergovernmental Pilot Project, 2016, p. 10). The intention was to overcome divisions imposed on the regional problem of SLR planning by local jurisdictional boundaries, sectoral boundaries, and the boundaries around federal properties and services. The mission of the Pilot Project was to "develop a regional 'whole-of-government' and 'whole-of-community' approach to sea level rise preparedness and resilience planning in Hampton Roads that also can be used as a template for other regions" and to put into place "intergovernmental planning organizational arrangements and procedures that can effectively coordinate the sea level rise preparedness and resilience planning" (Hampton Roads Sea Level Rise Preparedness and Resilience Intergovernmental Pilot Project, 2014, p. 1).

The driving force for the Pilot Project was the White House Executive Order (EO) 13653 on Preparing the United States for the Impact of Climate Change, issued on 1 November 2013. EO 13653 established the State, Local and Tribal Leaders Task Force on Climate Preparedness and Resilience. Neither Hampton Roads nor the Commonwealth of Virginia were represented on the Task Force. However, the White House later agreed to support a pilot project in Hampton Roads to address SLR resilience and coordination of these efforts across multiple jurisdictions.

The Pilot Project was formally announced in June 2014. Old Dominion University (ODU) agreed to a convening and facilitating role for the Pilot Project, supporting a Steering Committee that would lead the effort. In October 2014, the Steering Committee approved the charter, and working groups and advisory committees met regularly from December 2014, using the Pilot Project charter to guide their work. An interim report for Phase 1 was delivered in June 2015 (Hampton Roads Sea Level Rise Preparedness and Resilience Intergovernmental Pilot Project, 2015).

The working groups and advisory committees continued their work in Phase 2, focusing on strategies for collaboration, identifying case studies, and developing recommendations. The Steering Committee adopted a final resolution, recommending the HRPDC as the initial entity to provide continued leadership consistent with the conclusions of the Pilot Project. A final report for the Pilot Project was issued in October 2016 (Hampton Roads Sea Level Rise Preparedness and Resilience Intergovernmental Pilot Project, 2016).

The Pilot Project had numerous built-in organizational structures. In addition to a Steering Committee, five working groups were established, which were supported by three advisory committees intended to provide information and support (see Figure 3.2). Several working groups and advisory committees provided liaisons

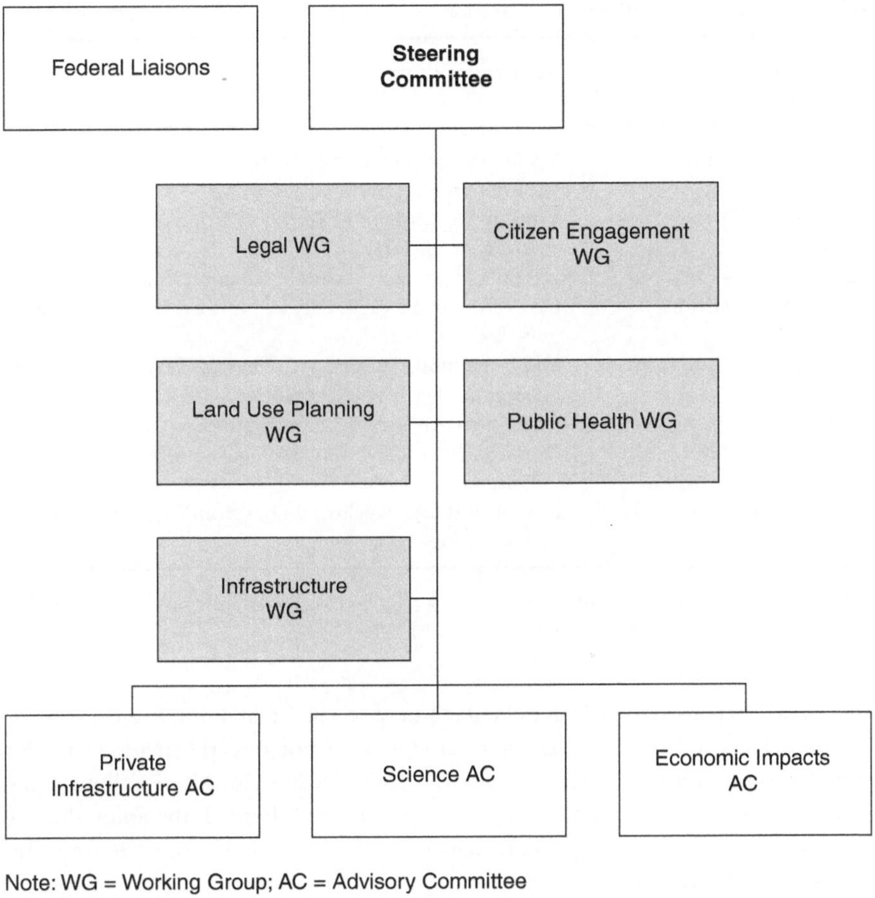

FIGURE 3.2 Structure of the Pilot Project
Source: Yusuf et al., this chapter

between committees and working groups to improve information exchange. For example, the Legal Working Group and Science Advisory Committee provided liaisons to both the Infrastructure Working Group and the Private Infrastructure Advisory Committee.

Participation among Steering Committee members, liaisons, and working group and advisory committee members was voluntary. There was substantial volunteer time devoted to the Pilot Project by over 200 individuals representing over 30 organizations (see Table 3.1 for examples of participating organizations). There was extensive government involvement in the Pilot Project, including local, state, and federal government agencies.

The Steering Committee consisted of representatives of the shipbuilding industry, state agencies, and regional and local governments (Hampton Roads Sea

TABLE 3.1 Examples of different organizations participating in the Pilot Project*

Organization Type (No. of Organizations)	*Representative Examples*
Federal agencies (15)	US Department of Transportation, Environmental Protection Agency
State agencies (7)	Virginia Department of Transportation, Office of the Attorney General
Regional entities (3)	HRPDC, Hampton Roads Sanitation District
Local governments (8)	City of Virginia Beach, City of Norfolk, City of Newport News
Educational institutions (8)	Old Dominion University, College of William & Mary, Virginia Beach Public Schools
Non-profits (11)	Lynnhaven River NOW, Southeast Care Coalition
Private industry (20)	Dominion Virginia Power, Sentara Norfolk General Hospital
Professional associations (5)	Hampton Roads Realtors Association, Virginia Maritime Association

* Note: Participation may range from attendance at a Pilot Project event to extensive involvement in the Steering Committee, working groups, or advisory committees.

Level Rise Preparedness and Resilience Intergovernmental Pilot Project, 2016). Representatives from the federal government were not directly included in the Steering Committee in compliance with the United States Constitution limitations on federal government powers at the sub-federal level. Instead, the Pilot Project structure included non-voting federal liaisons to the Steering Committee from the US Navy, US Army Corps of Engineers, and US Coast Guard.

The working groups were largely comprised of government representatives. The exception was the Citizen Engagement Working Group, whose members represented academic, non-profit, environmental, and community groups. The advisory committees were also more diverse and inclusive. The Science and Economic Impacts Advisory Committees had researchers and scientists from various academic institutions throughout the region. Finally, the Private Infrastructure Advisory Committee had greater representation of private sector interests, such as the utilities, real estate, and medical industries.

Methodology

Analysis of the Pilot Project relies on participant observation, document analysis, and key informant interviews. Members of the research team were embedded within the Pilot Project, in key roles such as chairing or serving on Pilot Project committees and working groups and providing planning and research support for Pilot Project events. We coupled the knowledge from this participant observation approach with analysis of documents such as the Pilot Project charter document and the Phase 1 and Phase 2 reports (Hampton Roads Sea Level Rise Preparedness and Resilience Intergovernmental Pilot Project, 2014; 2015; 2016).

We conducted interviews with 12 key informants, including members of the Steering Committee, federal liaisons to the Steering Committee, and chairs and members of working groups and advisory committees. Interviewees were selected to represent the range of local, regional, state, and federal government agencies and to ensure we incorporated the perspectives of the multiple sectors – government, business, non-profits, academic, and civil society. Structured interviews were conducted in spring 2017 to obtain perceptions about the Pilot Project objectives, outcomes, and success.

Findings

Findings can be organized into four major themes. First, there was a lack of clarity and consensus on how the Pilot Project would achieve its purpose and what the key deliverable would look like. Second, the Pilot Project achieved some progress in planned adaptation to SLR by creating mechanisms and processes for dialog, interactions, sharing of information, and building relationships among government entities. Third, the Pilot Project did result in several collaborative outcomes that advanced planned adaptation in the region. Finally, there were other challenges that interviewees perceived as preventing the Pilot Project from being effective in establishing an institutional arrangement supportive of a whole-of-government and whole-of-community planning for adaptation to SLR, including lack of resources, lack of leadership and ownership, and lack of broader representation.

Lack of clarity and consensus on objectives and outcomes

The purpose of the Pilot Project was generally understood to be establishing a whole-of-government and whole-of-community organizational framework to coordinate SLR preparedness and resilience planning (Hampton Roads Sea Level Rise Preparedness and Resilience Intergovernmental Pilot Project, 2015; 2016). Interviewees had similar understandings of the purpose of the Pilot Project. However, there were subtle differences in how interviewees described the purpose and varying perspectives on achieving the objectives of the Pilot Project.

The lack of clarity and consensus involved three different perspectives. Some interviewees viewed the Pilot Project as an opportunity to bring people together and build relationships. This view of the Pilot Project was more about interactions among participants. For example, interactions with federal agencies on issues related to SLR were often mentioned as outcomes. Representing a local government agency, one interviewee stated that the local governments generally did not understand what the federal agencies were doing (and could do) to plan for SLR. At the federal level, one interviewee mentioned that the Pilot Project helped participants by expanding their networks, creating new contacts, and developing deeper relationships. However, participants were in agreement that the Pilot Project was not successful in sustaining or institutionalizing these networks and relationships, with one noting that there was "no formal process in place to keep relationships and create more moving forward."

A second perspective among interviewees was that the Pilot Project initiated a process to develop a regional approach. Some interviewees echoed the wording of the Pilot Project mission, referring to a "framework" or "model" for collaboration. Specifically, one participant who represented business interests stated that the purpose of the Pilot Project was to develop a framework for regional collaboration in part by breaking down the barriers to collaboration. Another local government representative stated that the Pilot Project was an opportunity to "figure out how to address the SLR problem as a region." Third, interviewees viewed the Pilot Project as a multi-sectoral, intergovernmental, and collaborative effort to regionally address SLR, stressing the need for broad representation across governments and sectors. Several interviewees noted the need for diverse participants, specifically mentioning federal and community perspectives while expanding participation from the 17 local governments in the region.

These three perspectives illustrate the lack of clarity and consensus on Pilot Project objectives and outcomes, reflecting differences in degrees to which the Pilot Project was expected to address the who, how, and why of regional collaboration for SLR. The first view emphasizes learning about others involved in SLR planning, the *who*, while the second view is about developing a process for how to collaborate, the *how*. The third view focuses on the breadth of SLR as a regional challenge with many different stakeholders with multiple motives, the *why* for pursuing whole-of-government and whole-of-community collaborative efforts.

Creating mechanisms and processes for relationship building

Interviewees were consistent in expressing their approval of the Pilot Project's organizational structure, particularly at the working group level. When referring to their interactions within the working groups, several noted that they expanded their peer networks, built the foundation for respectful relationships, and initiated conversations among participants representing various sectors. However, interviewees voiced concern that momentum in the governmental sector might not be matched by similar engagement in the other sectors, particularly the civic community.

The unique mix of representatives in the working groups and advisory committees facilitated information sharing. In some cases, simply having the opportunity to work with the academic community, the private sector, non-profit organizations, and differing levels of government, enabled members to build relationships external to their normal networks, thus creating a unique policy network. During the Pilot Project, this policy network took the shape of a policy community, where participants share broad values, beliefs, and preferences, meet frequently, and interact closely on issues related to SLR and flooding adaptation. Interviewees revealed that the relationships that began during the Pilot Project had the potential to become the foundation for continued conversation and interaction towards achieving region-wide policies for adapting to SLR.

Although the concept of a whole-of-community involvement was an integral part of the Pilot Project's charter document, most of the focus was on policymaking at the local, regional, state, and federal levels. Getting "the community" (eg, homeowners, renters, faith-based organizations, non-profits, etc.) involved in a sustained and meaningful way was difficult to do, and there were questions raised about "the right time" to bring the public into the effort. One interviewee urged that both messaging and information needed to be consistent throughout the region and regularly transmitted to community members, which was a challenge given the lack of clarity and consistency about the intended message, as raised above.

Mixed success

As noted in the Phase 2 report, the Pilot Project's conclusion was "different than originally expected when the charter was developed," but that "the process in and of itself brought hundreds of stakeholders together, built lasting and ongoing relationships, and produced many workable recommendations for the region that can be accomplished by a variety of partnerships" (Hampton Roads Sea Level Rise Preparedness and Resilience Intergovernmental Pilot Project, 2016, p. 10). Interviewee assessments were generally mixed, and not always in agreement with the stated conclusion documented in the Pilot Project Phase 2 report. Participants were positive about the relationships created. For instance, one local official stressed the new communication with the Navy, which was critical because of the Navy's extensive geographic presence in the local jurisdiction. However, some interviewees were skeptical about the continuation of those relationships beyond the Pilot Project, which contributed to perceptions of the Pilot Project's mixed success.

The Pilot Project did produce concrete deliverables that interviewees perceived as beneficial outcomes. For example, the Legal Working Group produced a Legal Primer that detailed federal, state, and local laws and regulations related to planning for SLR. The Citizen Engagement Working Group developed a stakeholder engagement approach, called the Action-oriented Stakeholder Engagement for a Resilient Tomorrow (ASERT), which could be applied to engage residents in planning for flooding and SLR. The Infrastructure Working Group and the Private Infrastructure Advisory Committee conducted an assessment of vulnerable critical infrastructure in a selected watershed that spans multiple jurisdictional boundaries.

Despite these deliverables, some interviewees expressed that the Pilot Project largely missed out on meeting the stated objectives and outcome. The major shortfall was the lack of formalized regional collaboration involving the federal government, state government, local government, and business and non-profit communities (and in some cases, the residential community). However, other interviewees discussed several subsequent collaborative outcomes, such as the three recent Joint Land Use Studies (JLUS) in the region (Hampton Roads Planning District Commission, n.d.; Hampton Roads Planning District Commission, 2017), one of which is a first of its kind in that it considers SLR as an encroachment that threatens military readiness and the community's quality of life (Hampton Roads Sea Level Rise Preparedness and Resilience Intergovernmental Pilot Project, 2016). Another collaborative outcome was the creation of the Commonwealth Center for Recurrent Flooding Resiliency (CCRFR), established by the 2016 Virginia General Assembly through House Bill No. 903 ("An Act to designate the Commonwealth Center for Recurrent Flooding Resiliency jointly at Old Dominion University, the Virginia Institute of Marine Science, and The College of William and Mary," 2016). The CCRFR is intended to provide state agencies, localities, and others with scientific and technical support to advance recurrent flooding resiliency (Commonwealth Center for Recurrent Flooding Resiliency, 2017).

Other perceived challenges

The general consensus regarding the major obstacle to the Pilot Project's success was that it was naively optimistic. Developing a whole-of-government and whole-of-community approach to regional collaboration was a lofty goal. There was, in general, little funding to support the Pilot Project activities. Lack of other resources such as staff support and community connections hampered the achievement of Pilot Project objectives; one interviewee referred to the Pilot Project as an "unfunded mandate."

The perceived lack of leadership and ownership was another challenge that had an extensive ripple effect in terms of holding back regional collaboration. Several interviewees pointed to the lack of a strong champion who would push the Pilot Project forward. ODU played an active role in the Pilot Project, but this was in a voluntary and temporary capacity. Not surprisingly, a clear majority of interviewees indicated significant concerns about leadership in the next stage of regional planned adaptation. One interviewee noted, "My concern is that nobody is picking up the ball; somebody needs to take ownership now."

The Steering Committee was expected to provide leadership; however, there was a general perception that it did not fulfill this expectation. Some interviewees felt that there were no clear reasons for how and why some entities were selected for inclusion on the Steering Committee. In addition, not all Steering Committee members signed the Pilot Project charter. This lack of explicit connection to the purpose of the Pilot Project raised questions about the legitimacy of the Pilot Project and the Steering Committee's commitment to the effort.

The lack of broad representation in the Pilot Project extended beyond the Steering Committee; the vast majority of the region's 17 local governments were not involved in the Pilot Project, and only large industries were invited to participate. In addition, other community partners such as environmental non-profits, grassroots organizations, and residential and/or neighborhood groups were only minimally involved. Many interviewees noted the need for broader participation. One interviewee stated: "Everyone has to be at the table, everyone has to have a voice." Interestingly, there was some lack of agreement regarding who should be more involved. For example, while many agreed that local governments needed to be more engaged, there was some disagreement regarding who within the local governments should be involved (eg, chief administrative officers vs. technical staff).

Related to the issue of lack of representation was the conflict over duplicated efforts by parallel organizations. For example, when the Pilot Project was in its beginning stage, the HRPDC created a Coastal Resilience Committee. Most local governments in the region assigned senior administrators to serve on the Coastal Resilience Committee and committed funds for the HRPDC to add a coastal resiliency planner position. The Pilot Project was perceived by some to be competing with the Coastal Resilience Committee for local government participation.

This perceived duplication and competition created subsequent structural challenges for advancing the Pilot Project and regional collaborative efforts. Some interviewees saw the HRPDC, as the regional planning organization, as the natural leader for regional collaboration, but others disagreed. Importantly, the HRPDC expressed some resistance in adopting a post-Pilot Project leadership role. To exacerbate the situation, some actors from regional and local entities actively resisted the Pilot Project out of concerns regarding local government authority and autonomy, in a push against regional efforts in general.

Summary

The Pilot Project struggled with enabling participants to move from collaboration toward mutually agreed upon action. The scope of environmental change may not correspond well with human-made boundaries that are designed to serve the need for optimizing bureaucracy but not necessarily for anticipating the fluxes within socio-ecological systems. Such scale mismatch, as this case study shows, is not only apparent across institutions and the environment. Indeed, the analysis revealed that the Pilot Project's large, macro-level ambitions were often hampered by more meso-level concerns, including lack of clarity regarding the roles of the participants. Some participants saw the Pilot Project as vital for simply becoming acquainted with others engaged in SLR adaptation planning, while others saw its value as offering a process for collaboration, and still others derived value from simply learning why planned adaptation to SLR was important. Additionally, there was no identification of which organizations had the capacity to effect whole-of-community initiatives. As one local government interviewee noted, "there was a door open to peer in" to cross-organizational efforts to address SLR, but participants "didn't walk through it." Some interviewees remarked that too few municipalities were engaged in the Pilot Project, contributing to role confusion.

Concerns about stasis regarding progress toward action are not new, as several studies of sea level rise policy have identified persistent barriers to action (Moser and Ekstrom, 2010; Mozumder, Flugman, and Randhir, 2011; Yusuf and St. John III, 2017). One way to better bridge the mismatch between the extensive scope of the Pilot Project and the particularized concerns about a lack of clarity regarding institutional actors' roles and responsibilities is to tap where actors see a capacity for overcoming barriers. An earlier study of Hampton Roads actors found that respondents across multiple sectors and on a regional basis viewed political leadership and public support as strengths that can be leveraged (Yusuf and St. John III, 2017). Both facets indicate how governmental and non-governmental actors offer important assets in continued efforts to build regional resilience in a whole-of-government and whole-of-community way.

Many actors within the Pilot Project continue to have inter-organizational, cross-sectoral relationships. The next stage could be for political leadership to connect with community stakeholders to engage in discussion regarding crucial next steps to build a whole-of-community capacity to encounter SLR so as not only to survive but to thrive. Each group – political leaders and residents (i.e., the public) – is individually notable for being supportive of adaptation solutions and resilience. Interactions that cross sectors, as well as geographic boundaries, may be the next necessary step in moving towards a more authentic whole-of-community approach to SLR resilience.

The wicked problem of SLR still lacks a clear solution, and that solution is not likely to materialize with only one sector involved or as a result of a collaboration involving any two single sectors. As one local government interviewee stated, "water has no jurisdictional boundaries." The policies, decisions, and actions (or inactions) of one jurisdiction have a direct impact on other jurisdictions or entities, whether that "other" is a government, business, or a part of the civil society. SLR impacts will continue to plague the Hampton Roads region while awaiting a more comprehensive whole-of-community approach to SLR resilience.

Lessons learned

The Pilot Project illustrates the need for environmental policies, such as those related to planned adaptation to SLR, that adopt a multi-sectoral, multi-level governance framework while also highlighting the challenges of using this framework. The multi-sectoral, multi-level framework underpins effective environmental governance, which encompasses all governmental and non-governmental processes and actions by which "political, economic, and social actors regulate interactions between humans and nature" (Bevir, 2009, p. 78). Environmental governance is embedded within various vertical and horizontal relationships between organizations involved in formulating and implementing policy (Hill and Hupe, 2002).

In the case of regional environmental policy, horizontal linkages exist between political and social actors, between local governments, between departments and agencies within local governments, and so on. Vertically, linkages take place in the multi-level system of intergovernmental relations within which environmental policies are formulated and implemented. Pursuing environmental policies in pursuit of sustainability in a regional context requires successfully connecting multiple sectors and multiple levels of government and governance in a coherent and cohesive way. The Pilot Project highlights the challenges inherent in creating policy structures that encompass multiple levels.

As the product of bureaucratic-political struggles (Allison, 1971), policy can be ambiguous in its messages for implementation (Hill and Hupe, 2002). Inconsistent perceptions of policy outcomes underscore the political nature of the policies themselves (Barbeiro, 2014). Environmental policies can be perceived as having mixed success if the goals are vaguely defined (Layzer, 2012). This case study emphasizes the importance of clear specification of policy objectives and outcomes to avoid ambiguity (Hill and Hupe, 2002; Matland, 1995).

Findings complement an earlier study by Yusuf et al. (2016) of state-level agenda-setting identifying key factors contributing to SLR policy deficit, including the lack of clear consensus on policy solutions. Resource inadequacies also constrain environmental policies (such as those related to SLR) in Virginia (Yusuf et al., 2016), the Florida Keys (Mozumder et al., 2011), and Hong Kong (Francesch-Huidobro, 2012). The analysis of Virginia SLR policies (Yusuf et al., 2016) also squarely located SLR policy within a multi-level governance framework. The Pilot Project was responsive to this multi-level governance framework and incorporated a multi-sectoral approach, and was able to achieve some degree of the vertical integration that McGloughlin and Sweeney (2011) point to as being necessary for multi-level governance. The Pilot Project shows how difficult it is to integrate multiple sectors into the policy framework, yet the Pilot Project was not unique in terms of the challenges to environmental policies involving a multi-sectoral and multi-level governance framework. The lessons learned can likely be applicable to similarly situated environmental policies in pursuit of sustainability.

References

Adger, W. N., Hughes, T. P., Folke, C., Carpenter, S. R., and Rockström, J. (2005). Social-ecological resilience to coastal disasters. *Science*, 309(5737), 1036–1039.

Allison, G. T. (1971). *Essence of Decision: Explaining the Cuban Missile Crisis*. Boston: Little, Brown and Company.

An Act to designate the Commonwealth Center for Recurrent Flooding Resiliency jointly at Old Dominion University, the Virginia Institute of Marine Science, and The College of William and Mary (2016). Approved 11 March 2016.

Atkinson, L. P., Ezer, T., and Smith, E. (2013). Sea level rise and flooding risk in Virginia. *Sea Grant Law and Policy Journal*, 5(2), 3–14.

Babcock, J. (2009). Regionalism in Hampton Roads. Remarks of James F. Babcock to Leadership Hampton Roads, May 2003 (ed. 2009). fhrinc.org/publications/regionalism-in-hampton-roads [Accessed 16 November 2015].

Bacon, J. (2015). The Case for a Regional Approach to Economic Development. Bacon's Rebellion: Reinventing Virginia for the 21st Century, September 21, 2015. baconsrebellion.com/2015/09/an-economic-argument-for-a-regional-approach-to-economic-development.html [Accessed 16 November 2015].

Barbeiro, R. P. (2014). *The Politics of Public Policy*. Boston: Pearson.

Bevir, M. (2009). *Key Concepts in Governance*. Los Angeles, SAGE.

Borgström, S. T., Elmqvist, T., Angelstam, P., and Alfsen-Norodom, C. (2006). Scale mismatches in management of urban landscapes. *Ecology and Society*, 11(2), 16.

Bryson, J. M., Crosby, B. C., and Stone, M. M. (2006). The design and implementation of cross-sector collaborations: propositions from the literature. *Public Administration Review*, 66(s1), 44–55.

Commonwealth Center for Recurrent Flooding Resiliency (2017). About. Retrieved from www.floodingresiliency.org/about/ [Accessed 22 December 2017].

Cumming, G. S., Cumming, D. H., and Redman, C. L. (2006). Scale mismatches in social-ecological systems: causes, consequences, and solutions. *Ecology and Society*, 11(1), 14.

Dietz, T., Ostrom, E., and Stern, P. C. (2003). The struggle to govern the commons. *Science*, 302(5652), 1907–1912.

Ezer, T. and Atkinson, L. P. (2014). Accelerated flooding along the US East Coast: on the impact of sea-level rise, tides, storms, the Gulf Stream, and the North Atlantic oscillations. *Earth's Future*, 2(8), 362–382.

Federal Emergency Management Agency (2011). A whole community approach to emergency management: principles, themes, and pathways for action. Federal Emergency Management Agency. Retrieved from www.fema.gov/media-library-data/20130726-1813-25045-3330/whole_community_dec2011__2_.pdf [Accessed 21 December 2017].

Folke, C., Hahn, T., Olsson, P., and Norberg, J. (2005). Adaptive governance of social-ecological systems. *Annual Review of Environment and Resources*, 30, 441–473.

Francesch-Huidobro, M. (2012). Institutional deficit and lack of legitimacy: the challenges of climate change governance in Hong Kong. *Environmental Politics*, 21(5), 791–810.

Governor's Commission on Climate Change (2008). Final Report: A Climate Change Action Plan. Retrieved from www.adaptationclearinghouse.org/resources/virginia-governor-s-commission-on-climate-change-final-report-a-climate-change-action-plan.html [Accessed 21 December 2017].

Grundmann, R. (2007). Climate change and knowledge politics. *Environmental Politics*, 16(3), 414–432. doi:10.1080/09644010701251656.

Hamilton, L. C. (2011). Education, politics and opinions about climate change evidence for interaction effects. *Climatic Change*, 104(2), 231–242. doi:10.1007/s10584-010-9957-8.

Hampton Roads Planning District Commission (2010). Climate Change in Hampton Roads: Impacts and Stakeholder Involvement. Retrieved from www.deq.virginia.gov/Portals/0/DEQ/CoastalZoneManagement/task12-03-08.pdf [Accessed 21 December 2017].

Hampton Roads Planning District Commission (2012). Climate Change in Hampton Roads: Sea Level Rise in Hampton Roads, Virginia. Retrieved from www.hrpdcva.gov/uploads/docs/HRPDC_ClimateChangeReport2012_Full_Reduced.pdf [Accessed 21 December 2017].

Hampton Roads Planning District Commission (2017). Norfolk and Virginia Beach Joint Land Use Study – Fact Sheet #1. Retrieved from www.hrpdcva.gov/uploads/docs/HRPDC_JLUS_FactSheet%201.pdf [Accessed 21 December 2017].

Hampton Roads Planning District Commission (n.d.). Joint Land Use Studies. Retrieved from www.hrpdcva.gov/departments/joint-land-use-studies/ [Accessed 21 December 2017].

Hampton Roads Sea Level Rise Preparedness and Resilience Intergovernmental Pilot Project (2014). Hampton Roads Sea Level Rise Preparedness and Resilience Intergovernmental Planning Pilot Project Charter. Retrieved from https://digitalcommons.odu.edu/pilot-project_charter/1/ [Accessed 21 December 2017].

Hampton Roads Sea Level Rise Preparedness and Resilience Intergovernmental Pilot Project (2015). Phase 1 Report: Accomplishments and Lessons Learned. Retrieved from https://digitalcommons.odu.edu/hripp_reports/1/ [Accessed 21 December 2017].

Hampton Roads Sea Level Rise Preparedness and Resilience Intergovernmental Pilot Project (2016). Phase 2 Report: Recommendations, Accomplishments and Lessons Learned. Retrieved from https://digitalcommons.odu.edu/hripp_reports/2/ [Accessed 21 December 2017].

Hampton Roads Sea Level Rise Preparedness and Resilience Intergovernmental Pilot Project Legal Working Group (2015). Legal Primer. Retrieved from digitalcommons.odu.edu/hripp_reports/3 [Accessed 21 December 2017].

Hill, M. and Hupe, P. (2002). *Implementing Public Policy*. Thousand Oaks, CA: Sage.

Jentoft, S. and Chuenpagdee, R. (2009). Fisheries and coastal governance as a wicked problem. *Marine Policy*, 33(4), 553–560.

Kleinosky, L. R., Yarnal, B., and Fisher, A. (2007). Vulnerability of Hampton Roads, Virginia to storm-surge flooding and sea-level rise. *Natural Hazards*, 40, 43–70.

Kreuter, M.W., De Rosa, C., Howze, E. H., and Baldwin, G.T. (2004). Understanding wicked problems: a key to advancing environmental health promotion. *Health Education and Behavior*, 31(4), 441–454.

Lach, D., Rayner, S., and Ingram, H. (2005). Taming the waters: strategies to domesticate the wicked problems of water resource management. *International Journal of Water*, 3(1), 1–17.

Layzer, J. A. (2012). *The Environmental Case: Translating Values into Policy*, third edition. Washington, DC: CQ Press.

Lazarus, R. J. (2008). Super wicked problems and climate change: restraining the present to liberate the future. *Cornell L. Rev.*, 94, 1153.

Levin, K., Cashore, B., Bernstein, S., and Auld, G. (2012). Overcoming the tragedy of super wicked problems: Constraining our future selves to ameliorate global climate change. *Policy Sciences*, 45(2), 123–152.

Li, H., Lin, L., and Burks-Copes, K. A. (2012). Modeling of coastal inundation, storm surge, and relative sea-level rise at Naval Station Norfolk, Norfolk, Virginia, USA. *Journal of Coastal Research*, 29(1), 18–30.

Matland, R. E. (1995). Synthesizing the implementation literature: The ambiguity-conflict model of policy implementation. *Journal of Public Administration Research and Theory*, 5(2), 145–174.

McCright, A. M. and Dunlap, R. E. (2011). The politicization of climate change and polarization in the American public's views of global warming, 2001–2010. *Sociological Quarterly*, 52(2), 155–194.

McFarlane, B.J. (2013). Coastal Resiliency: Adapting to Climate Change in Hampton Roads. Retrieved from Chesapeake, VA: www.hrpdcva.gov/uploads/docs/07182013-PDC-E9I.pdf [Accessed 21 December 2017].

McGloughlin, J. S. and Sweeney, J. (2011). Multi-level climate policies in Ireland. *Irish Geography*, 44(1), 137–150.

Moser, S. C. and Ekstrom, J. A. (2010). A framework to diagnose barriers to climate change adaptation. *Proceedings of the National Academy of Sciences*, 107(51), 22026–22031.

Moser, S. C., Jeffress Williams, S., and Boesch, D. F. (2012). Wicked challenges at Land's End: Managing coastal vulnerability under climate change. *Annual Review of Environment and Resources*, 37, 51–78.

Mozumder, P., Flugman, E., and Randhir, T. (2011). Adaptation behavior in the face of global climate change: Survey responses from experts and decision makers serving the Florida Keys. *Ocean and Coastal Management*, 54(1), 37–44. doi:10.1016/j.ocecoaman.2010.10.008.

Pyke, C. R., Thomas, R., Porter, R. D., Hellmann, J. J., Dukes, J. S., Lodge, D. M., and Chavarria, G. (2008). Current practices and future opportunities for policy on climate change and invasive species. *Conservation Biology*, 22(3), 585–592.

Wu, S.-Y., Najjar, R., and Siewert, J. (2009). Potential impacts of sea-level rise on the Mid- and Upper-Atlantic Region of the United States. *Climatic Change*, 95(1–2), 121–138. doi:10.1007/s10584-008-9522-x.

Yusuf, J.-E. W. and St. John III, B. (2017). Stuck on options and implementation in Hampton Roads, Virginia: an integrated conceptual framework for linking adaptation capacity, readiness, and barriers. *Journal of Environmental Studies and Sciences* 7(3), 450–460.

Yusuf, J.-E., Neill, K., St. John III, Ash, I. K., and Mahar, K. (2016). The sea is rising… but not onto the policy agenda: A multiple streams approach to understanding sea level rise policies. *Environment and Planning C: Government and Policy*, 34(2), 228–243.

Yusuf, J.-E. W., St. John III, B., and Ash, I. K. (2014). The role of politics and proximity in sea level rise policy salience: a study of Virginia legislators' perceptions. *Journal of Environmental Studies and Sciences*, 4(3), 208–217.

PART II

Land management and land use

4

US PUBLIC LANDS AND A NEW ADMINISTRATION

New and old issues

John Freemuth

Federal lands in the United States are breathtaking in their ecological scope, their bountiful natural resources, and their policy complexity. The history that culminated in current policy arrangements for federal land management is long and convoluted. Today, four major federal land agencies manage about 27% of the United States landmass, much of which is concentrated in the West. The United States Forest Service (USFS) manages about 192 million acres, the National Park Service about 79 million acres, the United States Fish and Wildlife Service 89 million acres, and the Bureau of Land Management (BLM) 248 million acres (Congressional Research Service, 2017).

The federal land system was established over time. It came into being in what has been called the Reservation era of US land policy, when the decision was made to retain federal land in national public ownership for a variety of reasons. Most set the beginning of this policy era with the creation of Yellowstone National Park in 1872 (Clawson, 1983). Yet to some, that arrangement has never set easy. As early as 1912, the Chief of the Forest Service William Graves noted that:

> These are *public resources* which should be handled in the interests of the public. Moreover, the problems involved are such that they should definitely remain in the hands of the National rather than be turned over to the State governments. The property belongs to the *Nation* as a whole, and every citizen has an interest in it… The underlying purpose of the proposed *transfer* of the national forests to the States is really not to substitute State for Federal control, but rather to substitute individual for public control. Its most earnest advocates are the very interests which wish to secure such control… The proposition is one which the people as a whole would repudiate in an instant if they understood what is proposed.
>
> *Dana, 1956, p. 179, emphasis added*

Over a century after their initial beginnings, the future direction of the public lands and their management is high on the political agenda again. The occupation of the Malheur National Wildlife Refuge, the earlier Bundy affair in Nevada, and various calls for transfer giving the lands "back" to the states, and even privatization, have resounded throughout the American West and elsewhere. Efforts in collaboration appear at times to push back against this radical change in direction. Outgoing President Obama used the Antiquities Act to protect areas Congress would not, such as the Bears Ears in eastern Utah, and to broaden the scope of the American story through the designation of areas like Cesar Chavez's birthplace. Federal land agencies struggle with issues of diversity in ways that have become more visible than ever before. That struggle occurs while agencies try to adapt to demographic changes, retirements, and budget shortfalls that affect land management in myriad ways. If that were not enough, others seek well-intentioned though perhaps silver bullet policies of federal land management reform, some calling for a new Federal Land Law Review Commission (National Advisory Board, 2009). The Commission would be the first since the late 1960s, perhaps influenced in part by the "process gridlock" facing USFS (USFS, 2002), said to be caused by too many procedural requirements from laws such as the National Environmental Policy Act (NEPA) that lead to slow decision-making.

Reforming federal land policy is not, however, an altogether illegitimate proposal, as there have been a number of responsible propositions offered over the past 30 years. This chapter concentrates on two of many issues facing federal lands and federal land management: current policy reform ideas and controversy over the Antiquities Act.

Federal public lands: reform

> The federal Constitution forms a happy combination in this respect; the great and aggregate interests being referred to the national, the local and particular to the State legislatures.
>
> *Federalist 10*

The federal land system is managed (in almost all cases) by agencies of the national government. As noted, this arrangement has never rested easily with some individuals, and transfer proposals have continued from 1912 onward. In 1928, the national government offered to transfer the public domain (non-national park and national forest land and what would later become the land managed by the BLM) minus the mineral estate to the states. Secretary of the Interior Wilbur made the offer in Boise, Idaho, and President Hoover proposed it in Salt Lake City later the same year. Secretary Wilbur hinted that National Forests might eventually be considered, although Hoover did not. Echoing *Federalist 10*, Hoover did say that it was "not proposed to transfer forest, park, Indian, and other existing reservations which have a distinctly *national* as well as local importance" (Dana, 1956, p. 232). To Hoover, those lands apparently met James Madison's definition of belonging in the national interest and should remain under national management. The states, unhappy without the mineral estate, did not take up the offer.

The Sagebrush Rebellion of the mid to late 1970s is the most notable and familiar example of both rebellion and reform. McGreggor Cawley, in *Federal Land, Western Anger* (1993), suggested that Westerners supportive of traditional and historic resource use had become alarmed by a number of factors that led to this version of "rebellion." One issue was the redefinition of "conservation" by Secretary of the Interior Udall, which was a more preservation and resource protection oriented philosophy (Cawley, 1993, p. 24). The second factor was the passage of the Federal Land Policy and Management Act (FLPMA), which made it clear that lands managed by the BLM would remain under federal management, challenging a common Western view that the Taylor Grazing Act's reference to the public domain land's "final disposal" really meant to eventually "get rid of" (Cawley, 1993). That law read in part: "In order to promote the highest use of the public lands pending its *final disposal*" [emphasis added] (Public Law 73–482, 1934), hence the view that something fundamental had changed. Several Western states passed legislation claiming or asserting their right to at least the BLM land (Cawley 1993, p. 95, p. 109). The election of Ronald Reagan as President and his selection of James Watt, who was much more favorable to resource extraction, as Secretary of the Interior appeared to dampen the alarm, and the rebellion faded. This should be kept in mind as the discussion moves to the present day.

In 2012, the seemingly endless argument over what level of government ought to manage the federal land estate flared up again. Not surprisingly, it flared up in the West, led by individuals and groups in Utah and Arizona. In Arizona, the state legislature passed a bill that called for federal land agencies to give up title to roughly 48,000 square miles (124,000 square kilometers) of federal land by 2015. Arizona governor Jan Brewer vetoed the bill for reasons including cost and legal uncertainties. The Arizona legislature then placed Proposition 120 on the November ballot. The proposition called for the federal government to relinquish most non-Native American land within the state, including Grand Canyon National Park. Specifically, it stated that Arizona "Declares its sovereign and exclusive authority and jurisdiction over the air, water, public lands, minerals, wildlife and other natural resources within its boundaries" (Arizona, 2012). Arguments ranged from those touting state sovereignty through those asserting that federal land ownership was unconstitutional and included vague promises to protect "Grand Canyon State Park" to the degree the National Park Service had. The proposition failed by a vote of 67.7% to 32.3% (Arizona, Secretary of State, 2012). Nonetheless, the Arizona legislature recently brought this issue back again (Wildlife News, 2015).

In Utah, action was at the legislative level, with the passage of H.B. 148, called the "Transfer of Public Lands Act and Related Study." The law required the federal government to extinguish title to public lands and transfer title to those public lands to the state on or before 31 December 2014 (Utah, 2012, p. 79). The law mandated a study, "Report on Utah's Transfer of Public Lands Act H.B.148," which was prepared by Utah's Constitutional Defense Council and Utah's Public Lands

Policy Coordinating Office and presented to the 2013 general session of the legislature. The report contained several key sections, including historical background, management summaries, and proposed legislation creating a Public Lands Interim Commission to further study transfer. The report's governing assumption was that "those closest to, and whose lives are most directly impacted by, these public lands are better situated to make decisions regarding the use and enjoyment of these lands" (Utah, 2012, p. 4). By the end of the report, however, there was evidence that its authors realized that they were dealing with a complicated issue:

> The transfer of public lands contemplated by H.B. 148 is a bold initiative that will require a re-examination of public lands policy on a federal, state and county level. This re-examination must be fully informed and it implications thoughtfully evaluated. The many interests that have become institutionalized over the course of the past century must be identified, studied and given a voice in what must be characterized as a process. This process should have as its goal the development of a new vision for the public lands that better meets the economic, energy, education and recreation needs of today.
>
> *Utah, 2012, p. 79*

The Utah legislature began considering a group of bills that would set out how the state would manage the land. A potential controversy is readily apparent based on the comment of then House Speaker Betsy Lockhart: "The return on that investment in a win scenario is almost unbelievable in terms of the *trillions* of dollars over the next few decades that would be available to the people of Utah for our school system, for our infrastructure development, for everything. It's almost mind-boggling" (*Salt Lake Tribune*, 2014). Controversy over what would be done to the landscape in terms of resource development to raise that amount of money by developing the newly acquired lands was inevitable.

What is different this time, however, is that unlike every other Western protest about federal lands management, there does not appear to be a single focusing event like FLPMA mobilizing Westerners. Instead, many of the ideas seem to have come from far away, geographically and temporally, rather than from Westerners themselves. The American Legislative Exchange Council (ALEC) circulated a "Resolution demanding that Congress Convey Title of Federal Public Lands to the States" that was introduced in a number of Western states (ALEC, 2013). Only three members of the governing board of ALEC were actually from the West at that time.

There does appear to be a good deal of more nuanced opinion that the current federal land policy system is not working well. To restate, some of the opinions are likely grounded in the view that what might be termed the congressional promises of resource development, resource use, and multiple use have been broken. The setting aside of wilderness, national monuments, and the growing call for biodiversity and ecosystem protection are examples of a different policy direction, one that understandably makes a number of people upset, whether they be from local communities looking to recreate, local governments looking to bring in tourism, or extractive industries looking to profit.

Yet there is even wider agreement that the current policy approach of multiple use to most federal lands outside national parks and wilderness is not working. The multiple use doctrine has received a great deal of scholarly attention and criticism since 1960, when multiple use was codified in a new public land statute reflecting the governance of USFS land, the Multiple Use Sustained Yield Act (MUSY) of 1960 (Cawley, 1993; Nelson, 1995; Clawson, 1983; Cortner and Moote, 1998), describing "multiple use" land management as:

> Management of all the various renewable surface resources of the national forests so that they are utilized in the combination that will best meet the needs of the American people; making the most judicious use of the land for some or all of these resources or related services over areas large enough to provide sufficient latitude for periodic adjustments in use to conform to changing needs and conditions; that some land will be used for less than all of the resources; and harmonious and coordinated management of the various resources, each with the other, without impairment of the productivity of the land, with consideration being given to the relative values of the various resources, and not necessarily the combination of uses that will give the greatest dollar return or the greatest unit output.
>
> *US Code, 2017, Title 16, Sec. 531*

The included resources were outdoor recreation, range, timber, watershed, and wildlife and fish (Sec. 528). These were in addition to the original purposes of the national forests, which were "to improve and protect the forest within the boundaries, or for the purpose of securing favorable conditions of water flows, and to furnish a continuous supply of timber for the use and necessities of citizens of the United States" (US Code, 2017, Title 16, Sec. 475).

By the early 1990s, the multiple use issue had reached a critical mass and, because of FLPMA, involved the BLM lands too. In 1992, the House Interior Committee convened a workshop titled "Multiple Use and Sustained Yield: Changing Philosophies for Federal Land Management?" The report on that workshop provides a great deal of fodder on issues surrounding multiple use management and provided excellent foreshadowing of what was to gain momentum in years to come in terms of reform. As might be imagined, opinions varied as to the severity of the "crises" associated with multiple use; some speakers found that multiple use still had great potential. Yet it seemed that many more remarks were centered on reform ideas than on keeping the status quo. Those ideas ran the gamut from private public partnerships through free markets, public participation, and community governance, to ecosystem management (CRS, 1992). Since then, other ideas have proliferated. For example, Robert Nelson has written many thoughtful articles about public land reform, and in 2012, he argued, once again, that it was time to rethink federal land management:

> The solution is to be found in first identifying those areas of the public lands which are capable of yielding a national consensus as to their management purpose – likely to be the most environmentally attractive parts of the public lands, such as wilderness areas. The operative goals on the remaining, less nationally significant lands should be decentralization and privatization.
>
> *Nelson, 2012*

In 2014, Nelson, who is an economist with policy experience in the Department of the Interior, called for experimentation with a "charter forest" concept, much like charter schools. As noted in the *Wall Street Journal*:

> This is the model that the U.S. Forest Service needs. Certain federal forest lands, while still "owned" by the federal government, would be managed independently as charter forests. A decentralized charter forest would operate under the control of a local board of directors, which might include local government officials, economists, environmentalists, and recreational and commercial users of forest resources.
>
> *Nelson, 2014*

Nelson's charter forest is one of a group of ideas that revolve around localized governance arrangements. Some analyses have called for the establishment of forest or public lands trusts of varying sizes, and others argue for operationalizing the trust approach (maximizing revenue for beneficiaries like public schools) used in the management of many state land systems. The trust approach was tried for the Valles Caldera Preserve in northern New Mexico, beginning in 2000. However, that approach has failed and the area has reverted to the National Park Service for management (Fairfax and Souder, 1996; O'Toole, 2009; Valles Caldera Trust, 2014).

Collaborative groups, another form of local governance, have emerged throughout the West. What distinguishes them from the discussion above is that they are not the result of legislation or legal establishment. They have been sponsored at times by members of Congress; Senator Mike Crapo's sponsorship through legislation of the collaborative Owyhee Initiative in Idaho is one example (Owyhee Initiative, 2012). Sometimes they spring up through local discussions among various parties. Although in the future they might suggest certain public land policy reforms, today these groups usually work on finding agreement on issues like forest restoration timber harvests in specific localities (Clearwater Basin Collaborative, 2014).

Other proposed reforms focus more on alternative issues such as reforming user fees or what even the Forest Service has termed process gridlock resulting from too much contradictory legislation and complicated planning requirements. Finally, a number of observers have called for what might amount to a new Public Land Law Review Commission (PLLRC). The commission was established in 1964 to review then current public lands laws and to develop a public land policy. It issued a report in 1970 that led, among other things, to FLPMA. Cynics might be somewhat accurate if they called this a "full-employment act" for academics, lawyers, and federal land activists and managers. Such calls can also gloss over the politics that led to the creation of the PLLRC; a compromise between Congressman Wayne Aspinall of Colorado and the Kennedy/Johnson Administration that allowed for progress on the Wilderness Act (Cawley, 1993; Skillen, 2009). While admirable in intention, however, Congress has shown no interest in taking up such a proposal.

A policy case: the Antiquities Act

As a seasonal park ranger at Glen Canyon National Recreation Area in the 1970s, I hiked through several areas that were first protected as national monuments under the Antiquities Act, including Zion, Arches, Bryce, and Capitol Reef national parks, Natural Bridges National Monument, and the area that would become the Grand Staircase-Escalante National Monument. Much of the scenic Colorado Plateau region, which covers 140,000 square miles in the Four Corners region of the Southwest, has been protected from development under the Antiquities Act. While the Antiquities Act has played a crucial role in the growth of our national park system, it has become a flashpoint for disputes from Alaska to Maine over protection and use of public lands as it has been used to protect large areas of land that are not under the administration of the National Park Service. For that reason, it appears to work best when it is not used arbitrarily or too often and when the public understands and supports its use.

President Trump's Executive Order (April 26, 2017) to Interior Secretary Ryan Zinke initiating a review of certain national monuments designated by past presidents under the Antiquities Act likely had many people outside the Western US wondering what he was talking about and what it all meant. Words like "antiquities" and "monuments" may invoke images of old buildings, historical and symbolic structures like the Washington Monument, and ancient artifacts. The order, however, was much more than that. To understand why it has raised so much controversy, a review of the Act's history is important. Following that review, a survey of current issues follows.

The Antiquities Act in history

The Antiquities Act was passed by Congress in 1906, originally to conserve the stunning archaeological treasures of the American Southwest. As settlers, prospectors, ranchers, and explorers entered the region in the late 1800s, they discovered unique and spectacular sites left by Anasazi, ancestral Pueblo people who lived in the area from about AD 700 to 1600. These discoveries also led to looting and even shipping artifacts to institutions in the East and abroad. Concerned scholars and scientists began to call for controls. J. Walter Fewkes, a prominent archaeologist, warned in 1896 that unless this plundering of ancient sites was curbed, "many of the most interesting monuments of the prehistoric peoples of our Southwest will be little more than mounds of debris at the bases of the cliffs" (Harmon et al., 2006, p. 24). Spurred on by more vandalism and looting, Congress became involved.

With essential staff leadership by young archaeologist Edgar Lee Hewett and the sponsorship of Representative John Lacey (R-IA), Congress passed the "Act for the Preservation of American Antiquities" on 8 June 1906. The law's key provision states:

> That the President of the United States is hereby authorized, in his discretion, to declare by public proclamation historic landmarks, historic and prehistoric structures, and other objects of historic or scientific interest that are situated upon the lands owned or controlled by the Government of the United States to be national monuments, and may reserve as a part thereof parcels of land, the limits of which in all cases shall be confined to the smallest area compatible with proper care and management of the objects to be protected.
>
> *Public Law, 1906, 59–209, 34 Stat. 225, 54 U.S.C. § 320301–320303*

If this were all that happened, readers would be right to wonder what all the fuss regarding the Antiquities Act is about. Enter the Grand Canyon and President Theodore Roosevelt. How fortuitous it was for the national park system that Roosevelt – who strongly believed in using his authority to conserve precious natural resources – was President in 1906. Over three years Roosevelt designated 18 national monuments, some of which Congress later elevated to national parks or national historical parks, including Arizona's Petrified Forest National Park, a landscape of fossilized trees and Painted Desert; Chaco Culture National Historical Park, a stunning and complex Anasazi site in New Mexico; and Pinnacles National Park, a swath of rock spires and woodlands in California's Central Coast Range.

Roosevelt used the new Antiquities Act to set aside over 800,000 acres of the Grand Canyon, which was later designated as a national park by Congress. When his action was challenged in court as violating the "the smallest area compatible with proper care and management of the objects to be protected" clause of the Act, the court sustained his decision. In this key case, the Supreme Court ruled that the Antiquities Act gave President Roosevelt a broad power to, in the words of the Act, establish reserves embracing "objects of historic or scientific interest" (Cameron v. United States, 1920). Unsurprisingly, Roosevelt's argument was that the Grand Canyon, as was stated in the proclamation, "is an object of unusual scientific interest." Additionally, the Court did not consider the size of the monument within the proclamation to be too large. This ruling can be viewed as opening the door to other large monuments. Those monuments would come; some that later became national parks include Olympic, Zion, Bryce, Arches, Death Valley, and Joshua Tree.

In the 1930s, Wyoming residents objected when John D. Rockefeller offered to donate land that he owned near Jackson Hole to enlarge the original Grand Teton National Park. When Rockefeller threatened to sell the property instead, President Franklin Roosevelt combined the land with 179,000 acres from Grand Teton National Forest to create a national monument in 1943, which later was added to the national park. Wyoming Republican Senator Edward Robertson called the step a "foul, sneaking Pearl Harbor blow." In 1950, Congress amended the Antiquities

Act to require congressional approval for any future monuments designated in Wyoming (Harmon et al., 2006, chapter 5).

The next controversy flared in 1978, when President Jimmy Carter, acting on advice from Interior Secretary Cecil Andrus, designated 17 national monuments in Alaska totaling more than 50 million acres. Carter took this step after one of Alaska's senators, Mike Gravel, delayed action and threatened to filibuster pending legislation to create national parks, national forests, and wildlife refuges on these lands. Alaskans protested, but in 1980 Congress passed compromised legislation that converted the lands to parks and refuges. Once again, Congress amended the Antiquities Act to require congressional approval for any future national monuments larger than 5,000 acres in Alaska (Harmon et al., 2006, chapter 6).

In 1996, President Bill Clinton designated the Grand Staircase-Escalante National Monument in Utah, a spectacular swath of red rock canyons and mesas in the Colorado Plateau. Clinton administration officials sought to protect the areas from proposed coal mining nearby. The Interior Department tried to soften local opposition by offering Utah access to coal resources elsewhere through land exchanges, but Clinton proclaimed the monument without much advance consultation with local communities, leaving some Utahans feeling blindsided and resentful years later (Harmon et al., 2006, chapter 7).

Current issues

President Obama's December 2016 proclamation creating the Bears Ears National Monument, plus President Trump's recent Executive Order, have brought the Antiquities Act onto center-stage again. Trump's order had several specific review standards. Perhaps the most important standard is the size of a monument created by proclamation. His order states that:

> Secretary of the Interior (Secretary) shall conduct a review of all Presidential designations or expansions of designations under the Antiquities Act made since January 1, 1996, where the designation covers more than 100,000 acres, where the designation after expansion covers more than 100,000 acres, or where the Secretary determines that the designation or expansion was made without adequate public outreach and coordination with relevant stakeholders... to determine whether each designation or expansion conforms to the policy set forth in section 1 of this order. In making those determinations, the Secretary shall consider: [these are two key considerations]
>
> (i) The requirements and original objectives of the Act, including the Act's requirement that reservations of land not exceed "the smallest area compatible with the proper care and management of the objects to be protected."
> (ii) Whether designated lands are appropriately classified under the Act as "historic landmarks, historic and prehistoric structures, [or] other objects of historic or scientific interest" Presidential Executive Order on the Review of Designations Under the Antiquities Act. April 26, 2017.

The date of 1996 dovetails with the Clinton, Bush, and Obama presidencies, though the focus appears to be much more on the Clinton and Obama proclamations given the list of monuments. This order suggests President Trump can or will issue proclamations abolishing the monuments of Presidents Clinton, Bush, and Obama, and/or can or will make monument boundary adjustments as they have been made before. Abolishment by a president was only attempted once. In 1938, President Franklin Roosevelt considered abolishing the Castle-Pinckney National Monument in South Carolina. The Attorney General, Homer Cummings, was asked to rule on whether the President could do what he was contemplating. Cummings concluded Roosevelt could not abolish the monument. He argued: "The statute does not in terms authorize the President to abolish national monuments, and no other statute containing such authority has been suggested. If the President has such authority, therefore, it exists by implication" (Wyatt, 2016, p. 3). He also concluded that the President did not have implied authority to abolish national monuments. Roosevelt did not attempt to abolish Castle-Pinckney National Monument, but Congress abolished the monument later in 1951 (Wyatt, 2016, p. 3).

Further evidence can be found in FLPMA that expressly prohibits the Secretary of the Interior from modifying or revoking any withdrawal creating national monuments under the Antiquities Act. More interestingly, House Report 94–1163, which accompanied FLPMA, appears to infer that the prohibition may also apply to the President. This provision came with the House Bill, H.R. 13777, as introduced and as reported by the 94th Congress. The committee report stated:

> [The bill] would also specifically reserve to the Congress the authority to modify and revoke withdrawals for national monuments created under the Antiquities Act... These provisions will insure that the integrity of the great national resource management systems will remain under the control of the Congress.
>
> *Wyatt, 2016, p. 5*

One interpretation of this House report would suggest that Congress reserved the revocation power in terms of national monuments to itself. The report has not surfaced in any court cases to date, but one could assume it would if President Trump tried to revoke a national monument proclamation.

Additionally, it is worth noting that Congress also has designated, expanded, and modified boundaries of, and approved funding for, many national monuments, further complicating the legal ability of the President to rescind any monument. For example, Craters of the Moon National Park and Preserve is on the Order's review list, with a reported 737,000 acres, but this figure is incorrect for a number of reasons. First, it ignores the fact that the original Craters of the Moon National Monument was proclaimed by President Coolidge in 1926 at some 22,000 acres. There were a number of subsequent adjustments, with the largest coming with President Clinton's addition of some 660,000 acres to both the National Park Service and

BLM. But the list's acreage ignores the fact that Congress moved 410,000 acres of the NPS lands into a National Preserve. This puts all but 250,000 acres off limit to repeal by President Trump. However, in July 2017, Secretary Zinke recommended that the Craters boundary stay intact (Barnhill, 2017).

Perhaps of additional interest is Public Law 111–11, which established the National Landscape Conservation System (now the National Conservation Lands) under BLM. It includes all BLM national monuments, which are to be managed as a system, and the Secretary of the Interior must ensure that such management "protects the values for which the components of the system were designated." Does this law override any actions that President Trump might undertake? That remains to be seen.

Boundary changes

Can a president alter the boundaries of a national monument proclaimed by a predecessor? The answer is yes, because it has been done. One example involves Mount Olympus National Monument (now Olympic National Park), then under the management of the USFS, whose acreage was reduced by 313,280 acres by President Wilson to make Sitka spruce available for the war effort (National Park Service, 2017). Of course, the Forest Service had and continues to have a mission that allows for timber harvesting and, at the time, the National Park Service had yet to be established. However, the pertinent issue today links to the phrase in the Antiquities Act that calls for a monument to be "confined to the smallest area compatible with proper care and management of the objects to be protected," as this phrase is also used in President Trump's Executive Order calling for a review. The proclamation that creates a national monument is where the justification for the size of a monument will be found, based on what is being protected. What this review would then seem to demand is a study of the proclamation and its justification of the size of the monument based on the "historic landmarks, historic and prehistoric structures, and other objects of historic or scientific interest that are situated upon the lands owned or controlled by the Government of the United States." Arguably, a drastic reduction in size that did not have justification linked to the clauses above would be viewed as a surreptitious attempt to gut the monument itself and would almost certainly be challenged.

Conclusion

To some, presidential use of the Antiquities Act has become another example of executive over-reach. To others it is heralded as the vehicle that led to the protection of many of America's crown jewels: scenic landscapes, historic sites, cultural sites, and sites that remind us of both our mistakes and our successes as a nation. A case might be made that a more formal process of public consultation and involvement during the consideration of a monument designation might mollify the more thoughtful critics of the Act (Harmon et al., 2006, Chapter 8).

One can add to the Antiquities Act the fact that Congress once delegated to the president the power to proclaim forest reserves (now national forests) and that over 172 million acres on national forests in the American West are the result of actions of presidents, not Congress. One can also add that presidents, starting with Theodore Roosevelt, also created many of the early national wildlife refuges. If it were not for the actions of many of our presidents, the American public land system would be much smaller, poorer, and less renowned.

Another use of the Antiquities Act bears watching: using it to conserve and spotlight sites that mark important moments in American history, such as the new Cesar Chavez National Monument in California. Perhaps using the act to celebrate our history and learn from our failures can increase support for it as an enabling instrument of American conservation in all forms. As our population grows and our rich natural and historical heritage faces increasing threats, we should be looking to protect more places that can inspire and inform present and future generations and offer them recreational opportunities. That is the incomparable legacy of the Antiquities Act, as vital today as it ever was (Leshy and Squillace, 2017). It would be short-sighted in the extreme for Congress to change a single word of what has been, by practically every measure, one of the most fruitful and far-sighted environmental laws it has ever put on the books.

References

American Legislative Exchange Council (2013). Resolution Demanding that Congress Convey Title of Federal Public Lands to the States, [online] alec.org. Available at: www.alec.org/model-legislation/resolution-demanding-that-congress-convey-title-of-federal-public-lands-to-the-states/ [Accessed 29 January 2014].

Arizona (2012). Arizona's General Election Guide: Proposition 120, [online] azsos.gov. Available at: azsos.gov/election/2012/Info/PubPamphlet/english/Prop120.htm [Accessed 23 January 2013].

Arizona, Secretary of State (2012). Official Canvas, p. 18.

Barnhill, F. (2017). Craters of the Moon no longer up for review by Trump Administration, [online] boisestatepublicradio.org. Available at: boisestatepublicradio.org/post/craters-moon-no-longer-review-trump-administration#stream/0 [Accessed 14 July 2017].

Cameron v. United States (1920). 252 U.S. 450 (U.S. Supreme Court); 1.

Cawley, R. McGreggor (1993). *Federal Land, Western Anger.* Lawrence, Kansas: University of Kansas Press.

Clawson, M. (1983). *The Federal Lands Revisited.* Washington: Resources for the Future.

Clearwater Basin Collaborative (2014). Clearwater Basin Collaborative, [online] clearwaterbasincollaborative.org. Available at: www.clearwaterbasincollaborative.org/ [Accessed 29 January 2014].

Congressional Research Service (CRS) (1992). *Multiple Use and Sustained Yield: Changing Philosophies for Federal Land Management?* Prepared for the House Committee on Interior and Insular Affairs. Report No. 11.

Congressional Research Service (CRS) (2017). *Federal Land Ownership: Overview and Data.* Washington.

Cortner, H. and Moote, M. (1998). *The Politics of Ecosystem Management.* Washington D.C.: Island Press.

Dana, S. (1956). *Forest and Range Policy*. New York: McGraw Hill.

Fairfax, S. and Souder, J. (1996). *State Trust Lands: History, Management, and Sustainable Use.* Lawrence: University of Kansas.

Federal Land Policy and Management Act of 1976 (1976). Public Law, 94–579.

Harmon, D., McManamon, F., and Pitcaithley, T. (2006). *The Antiquities Act: A Century of American Archaeology, Historic Preservation, and Nature Conservation*. Phoenix: University of Arizona Press.

Leshy, J. and Squillace, M. (2017). The Endangered Antiquities Act. *New York Times*, [online] A23. Available at: www.nytimes.com/2017/03/31/opinion/the-endangered-antiquities-act.html?_r=0 [Accessed 3 April 2017].

National Advisory Board, Public Land and Resources Law Review School of Law, University of Montana (2009). *A Public Lands Agenda for the 21st Century*. Missoula, Montana. Available at: heinonline.org/HOL/LandingPage?handle=hein.journals/publan30&div=4&id=&page= [Accessed 2 June 2017].

National Park Service (2017). Olympic Historic Resource Study Appendix A: A Chronology of the public domain, [online] nps.gov. Available at: www.nps.gov/parkhistory/online_books/olym/hrs/appa.htm [Accessed 2 June 2017].

Nelson, R. (1995). *Public Lands and Private Rights: The Failure of Scientific Management.* Maryland: Rowman & Littlefield.

Nelson, R. (2012). Our Languishing Public Lands, Policy Review 171, Hoover Institute. Available at: www.hoover.org/publications/policy-review/article/106516 [Accessed 3 May 2012].

Nelson, R. (2014). Taking an ax to traditional forest management: The Charter School approach works for education, so why not apply it elsewhere? *The Wall Street Journal,* January 2. Personal copy.

O'Toole, R. (2009). *A Matter of Trust: Why Congress Should turn Federal Lands into Fiduciary Trusts*. Policy Analysis 630 Cato Institute.

Owyhee Initiative (2012). Available at: www.owyheeinitiative.org/index.html [Accessed 29 January 2014].

Public Law 59–209 (1906). 34 Stat. 225 An Act for the Preservation of American Antiquities.

Public Law 73–482 (1934). To stop injury to the public lands by preventing overgrazing and soil deterioration, to provide for their orderly use, improvement, and development, to stabilize the livestock industry dependent upon the public range, and for other purposes. Section 315.

Public Law 111–11 (2009). 123 Stat. 991.

Salt Lake Tribune (2014). House GOP making plans to manage now-federal land, [online] sltrib.com. Available at: www.sltrib.com/csp/cms/sites/sltrib/pages/printerfriendly.csp?id=57462645/ [Accessed 29 January 2014].

Skillen, J. (2009). *The Nation's Largest Landlord: The Bureau of Land Management in the American West*. Lawrence: University of Kansas.

Trump, D. (2017). Presidential Executive Order on the Review of Designations Under the Antiquities Act. April 26, 2017.

US Code (2017). Title 16, Sec. 475, 531.

US Forest Service (2002). *The Process Predicament: How Statutory, Regulatory and Administrative Factors Affect National Forest Management*. Washington.

Utah (2012). Report on Utah's Transfer of Public Lands Act H.B.148, November 14.

Valles Caldera Trust (2014). Available at: www.vallescaldera.gov/about/trust/ [Accessed 27 January 2014].

Wildlife News (2015). Available at: www.thewildlifenews.com/2015/03/18/legislative-moves-to-give-public-lands-to-the-states-advance-despite-little-evidence-of-public-support/ [Accessed 19 March 2015].

Wyatt, A. (2016). *Antiquities Act: Scope of Authority for Modification of National Monuments.* Congressional Research Service, November 14.

5

PERCEPTIONS OF CONTENTIOUSNESS

How individual traits shape environmental policy conflicts

Kristin Olofsson

The extraction of unconventional oil and gas using hydraulic fracturing is a complex environmental issue that has drawn contention from all branches of government in the United States – Congress, the White House, and numerous federal agencies – as well as state government and local municipalities. The Obama administration fought to regulate hydraulic fracturing on federal lands, an important step for national regulation. However, while many applauded the Obama administration's attempts to regulate, many others felt the efforts were shallow and did not go far enough (Harder and Gilbert, 2015). Regardless, Obama's hydraulic fracturing regulations were quickly reversed by the Trump administration (Tabuchi and Lipton, 2017). The Environmental Protection Agency (EPA) spent more than six years on a landmark study exploring the water-related impacts of hydraulic fracturing. When the agency's report was released in 2015, it was met with contempt from groups both opposed to and in support of hydraulic fracturing. The former felt the EPA's recommendations were not well founded and exposed citizens to potential danger, while the latter were disappointed in the lack of an enforceable directive eliminating future questions related to the environmental impacts of oil and gas extraction (DiChristopher, 2016). By December 2016, the EPA reversed course somewhat and highlighted the sometimes tenable but nonetheless real connection between oil and gas operations using hydraulic fracturing and contamination of drinking water (EPA, 2016). While oil and gas production utilizing hydraulic fracturing may have tapered off from its apex in 2015, the debate surrounding the extraction of unconventional oil and gas remains critical for environmental and energy policy (Heikkila et al., 2017).

The extraction of unconventional oil and gas using hydraulic fracturing is a hugely complex issue of land management and hits on many environmental issues including water supply and quality, air quality, emissions, public nuisance, energy security, economic impacts, energy costs, and public health matters. The issues involved are complex and require involvement from all levels of government, and consequently environmental policies in pursuit of sustainability are difficult to form. Understanding the policy process that underlies the development of environmental policy and the policy process in general requires understanding the individuals involved in environmental policymaking. The objective of this research is to understand the beliefs and policy preferences of individuals involved at the national level in the policies shaping the extraction of unconventional oil and gas in the United States, including individuals involved in policymaking, citizens impacted by hydraulic fracturing, individuals who work in hydraulic fracturing professionally, government regulators, environmental groups, interested media, and more. The research focuses on national-level policy actors as these individuals – in comparison with individuals active at only the state or local level – should be less susceptible to local-level variance. The purpose of this research is to understand how the individual-level beliefs and policy preferences of actors in environmental decision-making influence their views while participating in the policy process. By exploring the beliefs and preferences of individuals involved in a high-conflict policy issue, and how those belief structures shape perceptions, we can better understand the development of environmental policy conflicts. After a discussion of the case setting, the chapter first examines the characteristics of national-level policy actors involved in making policies regarding hydraulic fracturing and then explores how differences in characteristics among policy actors are associated with varying perceptions of issue contentiousness.

Case setting: hydraulic fracturing

In hydraulic fracturing, fluids and other chemicals are injected into wells at high volumes and pressure to fracture the rock and release previously "trapped" oil and gas for collection (Rahm, 2011). The technique of hydraulic fracturing is not new; its emergence dates as early as the 1940s (Hubbert and Willis, 1957; Phillips, 1972). The use of hydraulic fracturing increased dramatically around 2005 as advancements in horizontal drilling technologies opened up ways for the oil and gas industry to access shale oil and gas reserves that were previously not economically viable (Heikkila et al., 2014). As usage of the technique rapidly expanded, so too did the debate. Moratoriums and bans have been discussed in several US states, from New York and Pennsylvania to Oregon. Bans have been passed in three states to date – New York, Vermont, and Maryland, as recently as March 2017 – with the ban in New York in 2014 garnering significant national attention. Moratoriums and local-level bans have been contested in numerous highly visible cases in several state supreme courts.

The hydraulic fracturing policy issue is not only important for environmental and energy policy conflicts, but it is also critical to understanding the behavior of individuals involved in high-conflict policy issues. Hydraulic fracturing is marked by complex policy issues, interrelated and often inseparable motivations of policy actors, relevancy at multiple levels of government and regulatory agencies, and broad importance to the general public as an issue that is directly related to the economy and the daily lives of many citizens. The results of this study can help in drawing more general conclusions about high-conflict environmental policy issues. It is critical for policymakers and citizens alike to understand high-conflict environmental policy issues because there are implications for the quality of politics that emerge in these situations and related democratic and environmental justice concerns. In high-conflict policy issues, there may be a tendency towards polarization and stalemate, which is not productive for effective or quality policy. In terms of democratic concerns, we should study environmental policy conflicts because we learn more about who is participating and how, if there are barriers to participation, and whether there is an unjust distribution of benefits or unequal burden of risk. Issues related to hydraulic fracturing are fraught with concerns regarding environmental justice. The individuals directly impacted by hydraulic fracturing operations are not necessarily the same individuals who experience the benefits or who have the power and/or access to impact policy. While this research does not endeavor to address all questions related to hydraulic fracturing policy, it does contribute to understanding how individual beliefs and policy perceptions shape an environmental policy conflict.

The theoretical basis for the research is the Advocacy Coalition Framework (ACF). The ACF is a useful lens through which to view the development of high-conflict policy issues. According to the ACF, individuals are motivated to participate in policy activities and form coalitions to impact policy change owing to shared beliefs, in this case, regarding hydraulic fracturing. These individuals are assumed to be boundedly rational and to call upon pre-existing belief systems to process their opinions and perceptions of a policy issue. We can expect, therefore, that individuals will display shared beliefs and largely stable preferences regarding their policy positions and risk assessments. Importantly, individuals are motivated to form and maintain coalitions based on shared beliefs, and it is within these coalitions that policy actors endeavor to impact policy. This can become problematic for effective policymaking if coalitions become silos and perceptions of high-conflict issues become both polarized and stagnant. This research aims to interrogate the structures and perceptions of policy coalitions surrounding hydraulic fracturing in the United States to explore how these coalitions are impacting the policy process.

Data collection

Data collection started with a series of 11 in-depth interviews during autumn 2015 with policy experts in unconventional oil and gas extraction. The policy experts were from a range of organizations, including professional industry association representatives, federal agencies, non-profit environmental groups, advocacy groups, and government representatives. There was no geographic target for these interviews, but all interviewees were active in policy at the national level, as opposed to the state or local level, in matters concerning the extraction of oil and gas using hydraulic fracturing. The in-depth interviews were used to tailor the survey to reflect the shape of the issue at the national level, and the survey was then vetted by an advisory committee of policy experts and updated based on their recommendations. These interviews also contributed to the development of the sampling frame.

A survey of policy actors involved at the national level on issues related to the extraction of unconventional oil and gas using hydraulic fracturing was administered in the spring of 2016. A total of 468 policy actors were identified as actively involved in or knowledgeable about unconventional oil and gas development in the United States. Individuals were included without regard for organizational affiliation or profession. Rules for inclusion required that the individual had to demonstrate activity within the policy issue since 2008 and had to be currently active at the national level (as opposed to the state or local level). Similar sampling procedures have been used in past research at the state and local levels (Weible et al., 2016; Heikkila et al., 2017). The issue of hidden populations is a limitation of any sampling procedure in which the population is not known *a priori*, as is common in social science research (Heckathorn, 1997). Because there is no comprehensive register of policy actors, approximation methods must be used to identify the population of individuals that may be active on the issue and to review the composition of the final sample to explore possibilities of systemic errors in the creation of the survey sample. Individuals were first identified through the in-depth interviews and then used as seeds for online discovery of activity (Nohrstedt and Olofsson, 2016). For online discovery of activity, the internet is searched for any and all forms of activity as potential for inclusion: testifying at a public hearing; comments on a public forum; employment or involvement with private, public, or non-profit organizations active within the policy space; social media; traditional media; and many more. For example, if an individual with a federal agency mentions during the interview that she worked with a research group at a university, the website of the research group is found and then reviewed to potentially include any and all individuals within that group that are active within the policy issue of hydraulic fracturing at the national level. If perhaps the interviewee was connected to the university research group through a co-worker in another department, then that co-worker is identified for inclusion in the survey sample and the department's employees are all reviewed for potential inclusion as well.

The survey was administered via email using the Qualtrics platform. Of the 468 individuals who were administered the survey, 133 responded for a response rate of 28%, which is considered acceptable for online social science survey research (Nulty, 2008; Topp and Pawloski, 2002). The design of the survey was theoretically based on the ACF and drew on similar surveys administered in the same issue area (see summaries in Heikkila and Weible, 2013; Heikkila and Weible, 2015). Survey research is a common tool within the framework's empirical tradition. The survey was developed under the framework to uncover the beliefs, positions, and perceptions of policy actors involved in the issues surrounding hydraulic fracturing.

Results

There does not appear to be systematic bias in the organizational affiliation or position of respondents, limiting the potential for sampling bias. In comparison to the full sample, there was an equal spread of organizational affiliations among respondents. In terms of stance, it is nearly evenly distributed; respondents with a stance against hydraulic fracturing report that they would like to see hydraulic fracturing either stopped (16.8%) or limited (34.5%) in comparison with respondents with a stance in support of hydraulic fracturing that prefer production continue at the current rate (25.7%), expand moderately (14.6%), or expand extensively (8.4%). The average survey respondent holds at least a Bachelor's degree, tends toward moderate to slightly liberal political views, and describes their priority level regarding dealing with issues related to hydraulic fracturing in either a professional or personal capacity as "high" (see Table 5.1).

The average respondent has relatively high levels of experience in reading about issues related to hydraulic fracturing and slightly less experience researching and analyzing topics involved in hydraulic fracturing, with comparatively low levels of experience in planning, working, or managing oil and gas operations or living within visible proximity of oil and gas operations. Combined with the relatively high level of education within the sample, the variation in types of experience with issues surrounding hydraulic fracturing may indicate that this sample is more professionalized in their experience rather than experiencing the issue personally. Finally, the average respondent holds a position that oil and gas development that uses hydraulic fracturing should be limited somewhat or continued at the current rate, in comparison to other respondents who take more extreme positions that hydraulic fracturing should be either stopped altogether or expanded moderately or extensively.

There are some sizeable variations within these averages, however. Regarding education, the sample mode is for a Master's or professional degree, but that only represents one-third of the sample. Nearly as many individuals hold either a Bachelor's degree or a PhD or JD. While nearly half of all respondents reported that dealing with hydraulic fracturing issues is of high importance to their professional or personal goals, nearly one-third reported that it is only of low to moderate importance. The modal category of organizational affiliation of the respondent was Environmental or Conservation Groups, representing around 38% of the sample. This was followed by Consulting Firms or Think Tanks (15%), Federal Government (11%), Oil and Gas Industry (11%), Oil and Gas Professional Associations (7%), University Research (7%), Organized Citizen Groups (4%), State Government (2%), and Other (5%).

Respondent belief structures

We can now begin to explore the belief structures of these policy actors and associate those with our variable of interest: individual perceptions of the conflict,

TABLE 5.1 Survey questions

Please indicate what comes closest to your current position in relation to oil and gas development that uses hydraulic fracturing. It should be...

- Stopped
- Limited
- Continued at the current rate
- Expanded moderately
- Expanded extensively

Many political issues in a democracy can be characterized as contentious. Compared to other political issues in the United States, the level of political contention about oil and gas development using hydraulic fracturing in the United States is...

- Far less contentious
- Less contentious
- Just as contentious
- More contentious
- Far more contentious

Do the views and actions of those you disagree with on oil and gas development that uses hydraulic fracturing...

- Threaten you personally or professionally (eg, your job, values, income, or quality of life)?
- Threaten the United States?

How would you describe your working professional relationship with people you disagree with on the issue of oil and gas development that uses hydraulic fracturing in the United States?

- Not collegial
- Somewhat collegial
- Quite collegial
- Very collegial

How would you describe your working professional relationship with people you agree with on the issue of oil and gas development that uses hydraulic fracturing in the United States?

- Not collegial
- Somewhat collegial
- Quite collegial
- Very collegial

Over the past two years, has protection of the environment and public health in relation to oil and gas development that uses hydraulic fracturing in the United States become worse, stayed the same, or become better?

- Much worse
- Worse
- Stayed the same
- Better
- Much better

TABLE 5.1 *(cont.)*

When it comes to politics, do you usually consider yourself...

- Extremely liberal
- Liberal
- Moderate
- Conservative
- Extremely conservative

Please indicate the highest level of education that you have attained:

- Some college
- Bachelor's Degree
- Master's or Professional Degree
- JD
- PhD or MD

How much of a priority is it for you professionally or personally to deal with political and policy issues related to oil and gas development that uses hydraulic fracturing?

- Not a priority at all
- Low priority
- Moderate priority
- High priority
- Highest priority

Please indicate your level of experience with the following:

- Researching or conducting science on the technical aspects of oil and gas development
- Reading scientific studies about the economic, environmental, and public health impacts of oil and gas development
- Analyzing economic or financial impacts of oil and gas development
- Planning, working, or managing oil and gas operations
- Owning or leasing mineral or surface rights toward oil and gas development
- Living within visual proximity of oil and gas operations
- Regulating or governing oil and gas development
- Participating in political activities to influence government decisions about oil and gas development

measured as a respondent's opinion regarding the level of contentiousness of the policy issue of hydraulic fracturing. Respondents were asked to rate the level of political contention regarding oil and gas development using hydraulic fracturing, in comparison with other political issues in the United States. The Likert-type scale ran from 1 to 5 with the following response options: Far less contentious (1), Less contentious (2), Just as contentious (3), More contentious (4), Far more contentious (5). Table 5.2 summarizes the results.

The most common response was "Just as contentious," representing nearly 40% of the sample. The mean response for this question was 3.32 with a standard deviation of 0.94, indicating that generally, respondents find hydraulic fracturing to be at least as contentious if not more contentious than other political issues in the United

TABLE 5.2 Perceptions of level of issue contentiousness

	Frequency	*Percent*	*Cumulative Percent*
Far less contentious	3	2.80%	2.80%
Less contentious	16	14.95%	17.76%
Just as contentious	42	39.25%	57.01%
More contentious	36	33.64%	90.65%
Far more contentious	10	9.35%	100%
TOTAL	**107**	**100%**	**100%**

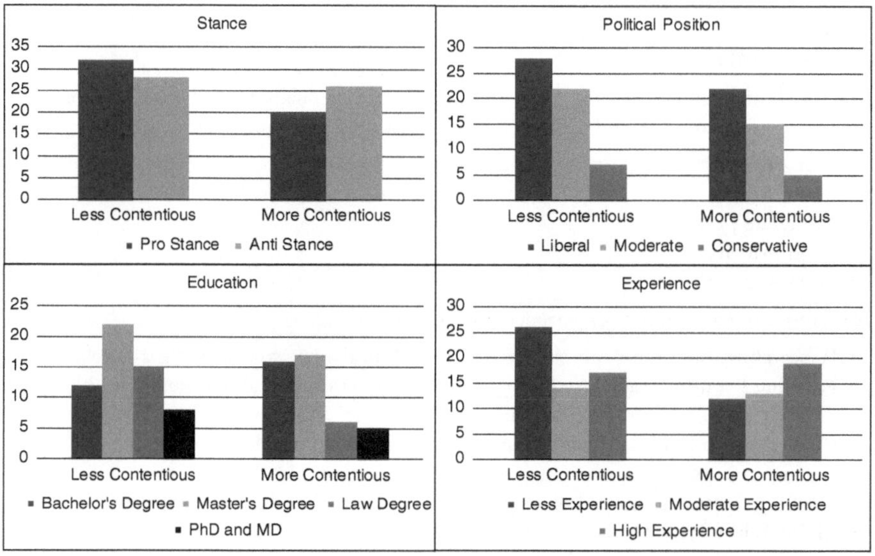

FIGURE 5.1 Individual traits by contentiousness
Source: Kristin Olofsson

States. Only about 18% of respondents felt it was less contentious or far less contentious. The standard deviation is relatively high in comparison to the mean, which indicates that there is fairly wide variation in the answers. However, that variation is concentrated within the group of respondents who feel the issue is at least as contentious or more so, which makes the distribution of responses largely irrelevant for our purposes. For ease of interpretation, responses on the perception of the level of issue contentiousness were collapsed into two categories: Less Contentious (n=61) and More Contentious (n=46). Figure 5.1 presents the individual traits discussed in the previous section, differentiated by perception of contentiousness.

It is encouraging that there are not strong patterns within the sample regarding stance, political position, and education in association with a respondent's perception of issue contentiousness. Strong patterns with regard to these basic individual traits such as education and stance might indicate sample bias or that the particular policy issue of hydraulic fracturing is defined along partisan lines. There appear to be some differences regarding experience associated with an individual's perception of the level of contentiousness of hydraulic fracturing, but it is important to note that none of the differences between and among groups were found to be statistically significant through two-way measures of association (Pearson's Chi^2 test).

In absolute terms, however, one of the more noticeable patterns is in professional and personal experience, which has also been collapsed into three categories for ease of interpretation. Individuals with less experience have a tendency to perceive hydraulic fracturing as less contentious, in comparison with individuals with moderate or high experience, who have a tendency towards a perception that the issue is more contentious. Caution is warranted in drawing generalizations regarding the association between experience and contentiousness, as this measure is rather blunt. It collapses several different types of experience, such as researching or analyzing the impacts of hydraulic fracturing, into one category with other types of experience, such as owning mineral rights or living in visual proximity of oil or gas operations.

To check for construct validity of the collapsed measure, all individual types of experience were tested individually against contentiousness in two-way measures of association using Pearson's Chi^2 test, and two were found to be statistically significant: "Regulating or governing oil and gas development" ($p<0.10$) and "Participating in political activities to influence government decisions about oil and gas development" ($p<0.05$). The omega-squared effect size for regulating experience indicates that this type of experience accounts for nearly 4% of the variance in perception of issue contentiousness, which appears relatively small but is still statistically significant. Even more pronounced is the omega-squared effect size for participating in political activities, which indicates that this type of experience accounts for more than 10% of the variance in perception of issue contentiousness. It appears that experience with particular types of activities does have an impact on an individual's perception of issue contentiousness.

We can now begin to explore the belief structures of individual respondents to understand how cognitive characteristics, such as perceived threat, may be associated with perceptions of issue contentiousness. Recent work on policy conflicts has shown that individual traits like education and stance are not as consistently influential upon perceptions of contentiousness as are cognitive characteristics such as threat perception or dissenting relationships (Weible and Heikkila, 2017; Heikkila and Weible, 2017). Figure 5.2 illustrates the associations between four cognitive characteristics and perception of the level of issue contentiousness.

All of the associations in Figure 5.2 were found to be statistically significant, indicating that there are non-random relationships between these cognitive characteristics and perception of issue contentiousness. Respondents were asked to gauge how the actions of those you disagree with on oil and gas development that uses hydraulic fracturing threaten both themselves and the United States. Response options were on a Likert-type five-point scale from "not at all" to "a great deal." The mean for personal threat was 2.57 and much higher for threatening the United

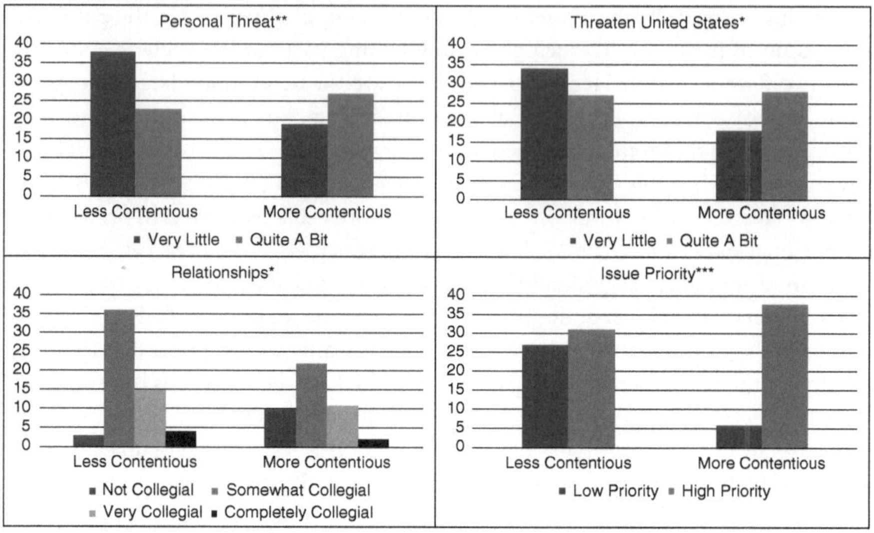

FIGURE 5.2 Cognitive characteristics by contentiousness
Note: Statistical significance: ***$p<0.01$, **$p<0.05$, *$p<0.10$

States at 3.40; however, when associated with issue contentiousness, differentiation is more pronounced for personal threat. Policy actors appear to associate threat and issue contentiousness: the less an individual feels threatened personally by the actions of those with whom they disagree, the less they feel that the issue of hydraulic fracturing is contentious, and vice versa. The same pattern holds for threat against the United States; an individual's threat tolerance is directly linked to their perception of issue contentiousness. The omega-squared score for personal threat is 0.081, indicating that perception of personal threat accounts for 8.1% of the variance in perceptions of issue contentiousness, while perception of threat towards the United States, although statistically significant, accounts for a much lower 1.8%.

Respondents were also asked to describe the collegiality of their relationships with people they disagree with on the issue of hydraulic fracturing. Response options were again on a Likert-type four-point scale from "not collegial at all" to "completely collegial." Average collegiality of relationships with those with whom a respondent disagrees was 2.23, around "somewhat collegial." Considering these are relationships among policy actors who are on opposite sides of an issue, a low mean is a valid result. The association between relationships with those with whom you disagree and contentiousness is slightly convoluted, but it is statistically significant. Respondents who take the extremes of collegiality, "not at all collegial" or "extremely collegial," are associated strongly with, respectively, "more contentious" and "less contentious" perceptions on the issue of hydraulic fracturing. This is an intuitive result that if a policy actor feels that their relationships are extremely collegial, even among those with whom they disagree, then it seems logical that the same policy actor would view the issue of hydraulic fracturing as less contentious.

The patterns among the moderate responses on relationships do not display an obvious trend. The collegiality of relationships among policy actors who disagree accounts for nearly 4% of the variance in perception of issue contentiousness.

In regard to issue priority, there are strong and statistically significant patterns associated with perceptions of issue contentiousness, but in ways that do not conform to patterns observed in previous respondent cognitive characteristics. Respondents were asked: "How much of a priority is it for you professionally or personally to deal with political and policy issues related to oil and gas development that uses hydraulic fracturing?" Five response options ranged from "not a priority" to "the highest priority." The mean was 3.69, above "a moderate priority" and close to "a high priority." The sample for this survey was relatively highly educated and more professionally than personally experienced on the issue of hydraulic fracturing, so this question may be picking up on that description. Nonetheless, there is a strong association that if an individual feels the policy issue of hydraulic fracturing is a high priority, they are also much more likely to perceive the issue as contentious, and the reverse holds for respondents who feel that the issue is of low priority.

Finally, respondents were asked to consider a measure of a policy outcome: protection of the environment and public health. Specifically, respondents were asked to gauge if protection of the environment and public health had become worse, stayed the same, or become better in the past two years. We might expect that those who believe that environmental and public health outcomes had deteriorated in the past two years would also perceive the issue to be more contentious. However, that expectation is not confirmed (see Figure 5.3). Although the association between

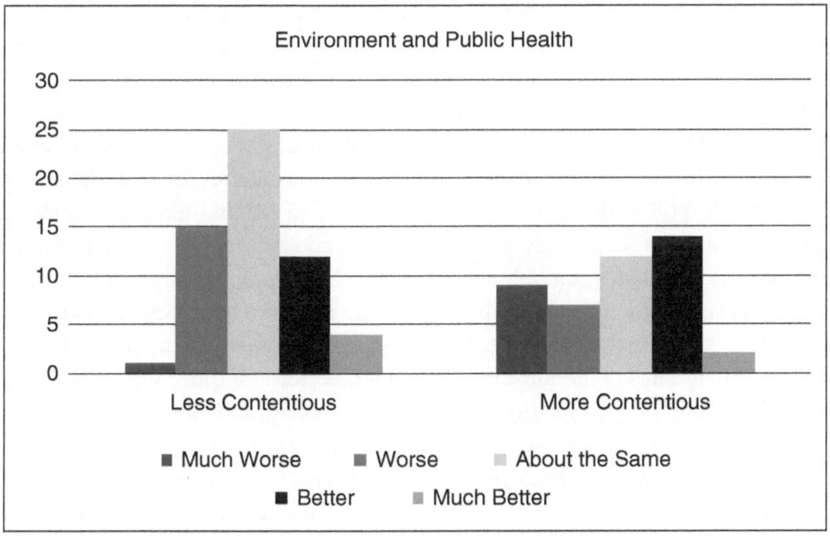

FIGURE 5.3 Environmental and public health outcomes by contentiousness
Source: Kristin Olofsson

environment and public health outcomes and issue contentiousness is strongly statistically significant, discernible patterns are not readily evident.

If an individual perceives outcomes to be much worse, and to some extent, even much better, then they are also more likely to perceive the issue as less contentious. It may be that policy actors who perceive deteriorating or improving conditions also feel that an issue may be less contentious as conditions shift from status quo and change occurs, for better or worse. The average for this question was –0.05 (on a scale from −2 for "much worse" to 2 for "much better"); the majority of respondents feel that environment and public health outcomes are about the same now as they were two years ago.

Conclusion

The research presented here reviewed an in-depth case study of the perceptions and priorities of individual policy actors involved in oil and gas development that uses hydraulic fracturing. The policy conflict surrounding oil and gas development that uses hydraulic fracturing is highly consequential for land use and management and is relatively contentious, and there are several individual traits and cognitive characteristics that are associated with this position. Policy actors feel that outcomes with regard to policies protecting the environment and public health have slightly worsened over the past two years. The issue is a priority for these policy actors, and they associate hydraulic fracturing with perceived threats, particularly personal threat perceptions. However, these policy actors have very little direct or personal experience with hydraulic fracturing and its impacts. In terms of general traits, there are not statistically significant differences in education, stance, political position, and experience – with some caveats – that appear to drive an individual's perception of issue contentiousness. Policy actors appear to be entrenched in their positions and threatened by the actions of those with whom they disagree.

For sustainable environmental policy, the implications of these patterns in cognitive characteristics are sizeable. Policy actor perceptions matter for making effective and sustainable policy that governs the extraction of oil and gas using hydraulic fracturing, while simultaneously protecting land and resources. Basic characteristics such as education, stance, or political ideology do not appear to be influential in determining issue contentiousness, but rather, the results of this study indicate that – in general – individual cognitive characteristics related to threat perception are more influential. Those who are threatened by the views of those with whom they disagree, have less collegial relationships with adversaries, and feel the issue is a high priority are also likely to be associated with a perception that hydraulic fracturing is a contentious issue. For policymakers and interested citizens, it appears to be important to moderate the levels of threat surrounding an issue and to encourage collegial relationships among opposing groups, in order to develop a policy area that is less contentious and, consequently, more amenable to long-lasting and equitable solutions.

There are limitations to this study. As discussed earlier, any sampling procedure that attempts to identify a population that is not known *a priori* will encounter problems of sampling bias. We attempted to correct for that limitation by vetting the sampling procedure and final sample frame with issue experts, who felt that our population adequately represented the spectrum of individuals involved. A related limitation is that of selection bias. Respondents who chose to answer the survey may over-represent extreme positions as they might be more motivated to share their opinions owing to their strongly held convictions on the issue. The rather moderate means on many questions would indicate that this is not an issue, but the risk remains and should be considered in drawing conclusions about a broader population of policy actors. Finally, any survey faces issues of survey instrumentality. The design and execution of a survey has been shown many times to be influential in responses. While we endeavor to present an unbiased survey instrument and to avoid all indications of preference, one respondent still felt that the survey was "misdirected" by taking up issues of hydraulic fracturing specifically when it should have been focused on shale oil more broadly. There is merit to this argument, as the policy issue of extraction of unconventional oil and gas has likely moved beyond just the technique of hydraulic fracturing, but the importance of studying this high-conflict policy issue remains high.

The potential impacts of hydraulic fracturing on land, water, air, and other natural resources are significant. Existing policy regulating hydraulic fracturing is fractured at best and lacks coherent best practices. Unlike some other environmental policies, the development of oil and gas extraction is an issue area that garners attention nationwide, from high-level law makers to everyday citizens. Those who live within proximity of industry operations inordinately bear the majority of the risks and harms associated with hydraulic fracturing, but the benefits are diffuse and enjoyed by the entire population of the United States, which presents significant policymaking challenges. The results of this study illuminate how individual traits and cognitive characteristics shape high-conflict policy issues. The data and analysis gathered here offer a rare insight into the personal belief structures of policy actors and how those belief structures shape issue perceptions. Theoretically, this study advances efforts to explain beliefs and preferences of policy actors involved in high-conflict policy issues. Practically, this study provides insight into how policy actors involved in high-conflict policy issues may differentiate themselves on perceptions of the level of contentiousness of a policy issue. Environmental policy is wrought with high-conflict issues, and it is imperative that we continue to explore how these issues take shape.

References

DiChristopher, T. (2016). Major EPA Fracking Study Downplayed Risks to US Water Supply, Investigation Finds. CNBC Broadcasting Corporation.

EPA (2016). EPA Releases Final Report on Impacts from Hydraulic Fracturing on Drinking Water, Washington, D.C.: Environmental Protection Agency.

Harder, A. and Gilbert, D. (2015). Fracking rules unveiled by Obama Administration. *Wall Street Journal*, March 20.

Heckathorn, D. D. (1997). Respondent-driven sampling: a new approach to the study of hidden populations. *Social Problems*, 44(2), pp. 174–199.

Heikkila, T. and Weible, C. M. (2013). *A Summary Report of Perceptions of the Politics and Regulation of Hydraulic Fracturing in Colorado*. Denver, CO: University of Colorado Denver, School of Public Affairs.

Heikkila, T. and Weible, C. M. (2015). *A Summary Report of a 2015 Survey of the Politics of Oil and Gas Development Using Hydraulic Fracturing in Colorado*. Denver, CO: University of Colorado Denver, School of Public Affairs.

Heikkila, T. and Weible, C. M. (2017). Unpacking the intensity of policy conflict: a study of Colorado's oil and gas subsystem. *Policy Sciences*, March.

Heikkila, T., Pierce, J. J., Gallaher, S., Kagan, J., Crow, D. A., and Weible, C. M. (2014). Understanding a period of policy change: the case of hydraulic fracturing disclosure policy in Colorado. *Review of Policy Research*, 31(2), pp. 65–87.

Heikkila, T., Weible, C. M., and Olofsson, K. (2017). Lessons from state-level and national-level policy conflicts over U.S. shale development. *Environment: Science and Policy for Sustainable Development*, 59(3), pp. 4–13.

Hubbert, M. and Willis, D. G. (1957). Mechanics of hydraulic fracturing. In *Transactions of Society of Petroleum Engineers of AIME*, pp. 239–257.

Nohrstedt, D. and Olofsson, K. L. (2016). Advocacy politics and strategies on hydraulic fracturing in Sweden. In C. M. Weible et al., eds. *Policy Debates on Hydraulic Fracturing*. London: Palgrave Macmillan, pp. 147–175.

Nulty, D. (2008). The adequacy of response rates to online and paper surveys: what can be done? *Assessment and Evaluation in Higher Education*, 33(3), pp. 301–314.

Phillips, W. J. (1972). Hydraulic fracturing and mineralization. *Journal of the Geological Society*, 128(4), pp. 337–359.

Rahm, D. (2011). Regulating hydraulic fracturing in shale gas plays: the case of Texas. *Energy Policy*, 39(5), pp. 2974–2981.

Tabuchi, H. and Lipton, E. (2017). How rollbacks at Scott Pruitt's E.P.A. are a boon to oil and gas. *New York Times*.

Topp, N. W. and Pawloski, B. (2002). Online data collection. *Journal of Science Education and Technology*, 11(2), pp. 173–178.

Weible, C. M., Olofsson, K. L., Costie, D. P., Katz, J. M., and Heikkila, T. (2016). Enhancing precision and clarity in the study of policy narratives: an analysis of climate and air issues in Delhi, India. *Review of Policy Research*, 33(4), pp. 420–441.

Weible, C. M. and Heikkila, T. (2017). Policy conflict framework. *Policy Sciences*, 50(1), pp. 23–40.

6

SOLID WASTE GOVERNANCE

Consumption and culture in the globalization era

Beatriz Adriana Venegas Sahagún

Proper urban waste management requires the participation of public and private organizations, civil society, and educational institutions. However, as the largest part of urban waste is generated by households, producing more waste than open-air markets, industry, parks and gardens, shopping malls, and others (Bernache et al., 1998), residential spaces play a critical role in urban waste management. Therefore, although municipal governments can put in place the infrastructure and regulations to facilitate and encourage waste management, waste segregation at the household level is necessary for such programs to be effective (Dai et al., 2015). Additionally, citizen participation in waste segregation also reduces the cost of urban waste management (Miranda Carreno and Blanco Suarez, 2010). Consequently, effective governance of urban waste management at the municipal level requires active citizen participation.

To encourage citizen participation, it is critical to determine what motivates or restricts residential members from participating in municipal solid waste (MSW) management (Hornik et al., 1995; Miafodzyeva and Brandt, 2013; Schultz, Oskamp, and Mainieri, 1995; Timlett and Williams, 2011). The work presented in this chapter shows that a stand-alone policy obligating municipal residents to segregate their waste may not be enough to motivate participation in urban waste management, and finds that other factors like socio-economic conditions, consumption habits, culture, and lack of relevant policy affect MSW management. The purpose of this chapter is to show how socio-political contexts and such context-dependent human behavior impact policy implementation and success, critical criteria in the pursuit of sustainability. As a solution, this study also explores the role of environmental education to make citizens aware of their urban waste management responsibilities.

The results presented in this chapter were gathered using mixed methods, including qualitative methods like ethnography, interviews, and documentary information, and quantitative methods via surveys. Residential members of the two communities were the main study participants. Additionally, the perceptions of other actors including waste collectors, scavengers, government officials, members of NGOs, and academics were also included to build a better understanding of the issues central to the research. The following section provides the conceptual foundation of this work and explains why consumption culture and behavior are important factors in effective urban municipal waste management. Then a brief background of the Mexican policies around MSW relevant to the municipalities chosen for this study is provided before turning to the specific case study.

Globalization and consumption

A part of the material stream that ends up as urban MSW does not begin at the residential unit but further away at the manufacturing stages, where a significant amount of packaging is used to facilitate the process of selling and transporting goods to consumers buying the goods in distant markets. Therefore, when these packaging materials reach the consumer, their purpose is served, and they directly end up in the waste stream (Bernache et al., 1998). As globalization has increased the consumption of products manufactured in distant lands, packaging materials in urban MSW have also increased.

Moreover, current consumption patterns create constant demand for a new variant of an existing type of commodity. Explaining this material consumption behavior, Appadurai (1991) points out that "modern consumers are victims of fashion" where fashion suggests "great speed, rapid turnover, illusion of full access and high convertibility, and presupposition of a democracy of consumers and objects of consumption" (p. 50). Such a lifestyle correlates with the increase in social metabolism and the increase of capital accumulation (Delgado Ramos, 2012), where the transformation of nature into goods or products reflects the increased accumulated capital of the society that consumes them. As a result, consumptive goods become obsolete based on fad rather than function, ultimately ending up in the urban solid waste stream, often as toxic waste (Stillerman, 2015). This has further aggravated problems of solid waste management in cities, as more and more usable but unfashionable products have flooded the waste stream (Bernache et al., 1998).

Additionally, it is not always up to individual consumers to decide whether to consume a certain good. According to Alonso (2005), currently we live in a "mass consumer society, where the idealization of the Western capitalist countries became a guiding axis of our daily life" (p. 8). Baudrillard explains it further by pointing out that people consume as "consumption is a mode of active relationship (not only with objects but also with the collective and the world)" (1979, p. 223). Even though consumption indeed existed in earlier times, it was not until the modern era of capitalism and subsequent globalization that people consumed not only for individual needs in response to the abundance of goods and services available to consume, but also to establish themselves within their peer groups by following group norms around the purchase of possessions and services (Belk, 2004).

Increased consumption is also linked with the purchasing power that comes with the economic power enjoyed by the economic power elites (McCracken, 1990). Economic power is "in the first place a power to put economic need at a distance: that is why it is universally affirmed by the destruction of wealth, ostentatious expenditure, waste and all forms of free luxury" (Bourdieu, 1988, p. 52). According to Bourdieu, purchasing power is shown as "the popular classes have as their principle the choice of what is necessary" (1988, p. 386), while "the ruling class (liberal professions or higher cadres) receive much and consume a lot" (p. 113). Not only that, Appadurai suggests that consumption is "an aspect of the total economy of societies. This means, demand arises as a function of the diversity of social practices and classifications" (1991, p. 47).

Societal pressure, increase in purchasing power, and the plethora of choices available in a consumer-driven society have increased consumption and ultimately garbage generation. The production of garbage is one of the main elements in environmental pollution, creating environmental fragilities and capabilities of destroying the natural environment (Ayres and Kneese, 1969; Bifani, 1999; Carson, 1962; Delgado Huertos, 1997). Therefore, it is critical to understand consumption behavior to find solutions to issues of MSW management. Additionally, it is also critical to understand the challenges of citizen participation in urban waste management, and what can influence effective urban waste management.

Policy context

Though Article 4 of the Mexican Constitution guarantees the right of every Mexican to live in an environment that adequately aids human development and well being, the beginning of environmental policies dates back to 1971 when the first federal environmental law was enacted in the country, named the Federal Law to Prevent and Control Environmental Pollution. The General Law on Ecological Equilibrium and Environmental Protection (LGEEPA) of 1988 resulted from this first legislative initiative in 1971, which in later years became foundational in paving the way for successive environmental legislation in the country. The LGEEPA also recognized the role and obligations of municipalities in environmental matters, especially in matters related to MSW management.

According to the LGEEPA, waste generated in Mexico is classified into three categories, and waste management is undertaken by the three levels of government: a) Hazardous Waste, managed by the federal government; b) Special Waste Management, controlled by the state governments; and c) Urban Solid Waste, managed by the municipalities. Therefore, MSW management including collection, transfer, and final disposal lies with local municipalities. Under the LGEEPA, progress was also made in the design and implementation of environmental policy using instruments at the national, regional, and local levels that recognized citizens' rights and stated their obligations. Therefore, LGEEPA not only defined the responsibility of the state and the different levels of government in waste management, it also allocated some responsibilities to the public to participate in their local environmental management activities (Rodríguez Becerra and Gabaldón, 2002).

The role of the local municipal governments in urban waste management was further defined by the constitutional reforms of Article 115 in 1983 and 1999, and in the General Law for the Prevention and Integral Management of Waste (LGPGIR) enacted in 2003, which introduced a new vision or model for public management in Mexico (Couto and Hernández, 2012). The first change to the constitutional Article 115 in 1983 established a municipal treasury of its own, allocations on the planning of the territory and the provision of public services that affect the quality of life of the inhabitant, and MSW management was tasked to municipalities. The second substantial change was in 1999 and defined the municipality as a governing body that strengthened the regulatory capacity of municipal governments (Acedo, 2007). Both constitutional changes changed the way governments make their own decisions and recognized the municipalities as local governments. Additionally, Article 115 of the Mexican Constitution enables the municipal governments to use different tools like direct benefit, concession, contract and sub-contract, agreement with the state, intergovernmental agreements, decentralized agencies, paramunicipal companies, escrow, and intermunicipal association to provide public services (García Del Castillo, 2003).

Different Mexican states enacted different environmental protection laws after LGEEPA. In the state of Jalisco, three laws that resulted from the LGEEPA related to solid waste management: 1) the State Law of Ecological Balance and Protection to the Environment of 1989; 2) the Law of Government and the Municipal Public Administration of 2000; and 3) the Law of Integral Management of the Residues of the State of Jalisco of 2007. In addition to those three laws, the state of Jalisco has the Norm NAE-SEMADES-007/2008, which established the criteria and technical specifications under which the separation, classification, selective collection, and recovery of waste in the state should be done. According to the state's laws, each municipality should have their own regulations in order to manage the MSW. The municipality of Zapopan has two regulations: 1) Regulation for the Management of Solid Waste of the Municipality of Zapopan, Jalisco, enacted in 2006; 2) Regulation of Protection to the Environment and Ecological Balance for the Municipality of Zapopan, Jalisco, enacted in 2011. The municipality of Tlaquepaque has one regulation: The Municipal Regulation of Ecological Balance and Protection to the Environment.

The legal framework for MSW and the attributions of municipal governments in MSW matters should lead to a good process of environmental governance because the functions and activities of the federal, state, and municipal government are defined in the laws, as are the obligations and responsibilities of and to the population, collecting companies, and other actors involved in the MSW management process. However, there is much room for improvement in the actual processes of MSW and disposal, including both citizen participation and the governance and management of waste. This chapter aims to explore the current context of MSW in order to offer suggestions for improvement to MSW governance.

Description of the study areas

Two municipalities in the Metropolitan Area of Guadalajara (AMG) of Zapopan and San Pedro Tlaquepaque in the state of Jalisco were identified for the research, following a comparative case study design (see Figure 6.1). The AMG is one of the most populated and economically developed regions in Mexico. The two communities were selected as they have very different management systems for MSW. The municipality of Zapopan has its own waste management system in place, with personnel and infrastructure to carry out the collection, sweeping, transfer, and final disposal activities. San Pedro Tlaquepaque, on the other hand, has mixed MSW management, where the municipality performs the collection, sweeping, and transfer to a privately owned plant in charge of the transfer and the final disposal.

The municipality of Zapopan is the second most populated of the AMG, after the municipality of Guadalajara. Zapopan has grown in recent years and has a population of 1,243,756 inhabitants (INEGI, 2010). The MSW generation of the municipality is 1,375,000 kg/day (3,031,356 lbs/day), representing an approximate average of 1.10 kg/person/day (2.42 lbs/person/day) of waste (INEGI, 2012). Traditional forms of economic activities like agricultural and livestock rearing has declined in the area, with the current focus mostly on manufacturing industries,

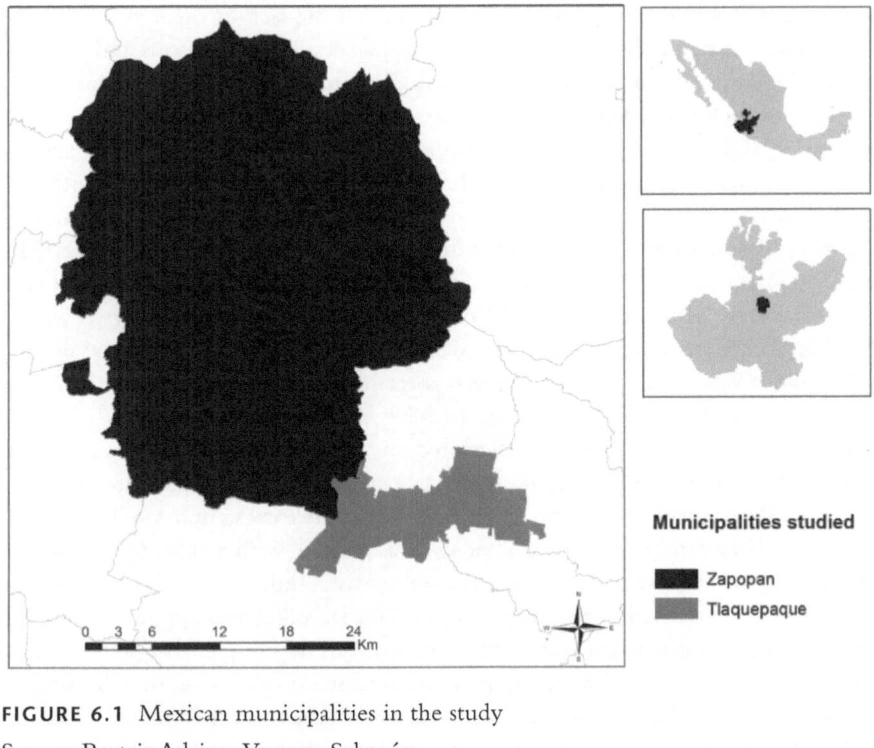

FIGURE 6.1 Mexican municipalities in the study
Source: Beatriz Adriana Venegas Sahagún

retail trade activities, temporary accommodation services, and food and beverage production activities (INEGI, 2010). The municipality of San Pedro Tlaquepaque has a population of 608,114 inhabitants (INEGI, 2010). In 2012, this municipality generated 467,570 kg/day (1,030,815 lbs/day) of MSW (INEGI, 2012), representing an approximate average of 0.77 kg/person/day (1.70 lbs/person/day). The main economic activities in this municipality involve manufacturing industries mostly related to producing handicrafts out of materials like papier-mâché, glass, brass, pottery, yarn, clay, leather, and wood (INEGI, 2010). Even though the economic activities in the two municipalities are different, the levels of waste generated by the two municipalities suggest similar levels of consumption. The two municipalities also have staff hired directly by the city councils for trash collection. Usually, two people are allocated per collection truck and other volunteers are recruited on a contractual basis by the municipality, with incomes dependent only on tips and the material they sell.

Research methods

Mixed methods were used to gather data for the research, using ethnographic research for qualitative data and surveys for quantitative data. An ethnographic study was carried out in the routes of the collection trucks of the two municipalities, in neighborhoods with high, medium, and low marginalization, as "marginalization is associated with the lack of social opportunities and the lack of capacity to acquire or generate them, but also to deprivation and inaccessibility of goods and services that are fundamental to well-being. Consequently, marginalized communities face scenarios of high social vulnerability whose mitigation is beyond personal or family control" (CONAPO, 2013, p. 11). This activity involved analysis of the practices of disposal, including the practices of the collector and their interaction with the society. Interviews were conducted with political actors in the municipalities as well as NGOs, educational institutions, and trade unions. Informal talks were held with the inhabitants of these municipalities, and the Ministry of Environment and Territorial Development (SEMADET) of the state of Jalisco organized two workshops on MSW management; the first workshop was about environmental governance of MSW and the second one was to elaborate the State Program of Waste Management in the State of Jalisco 2016–2022.

Also, a total of 385 people participated in a survey in the municipalities of Zapopan and San Pedro Tlaquepaque. Survey participants were approached in public spaces, like the parks or on the street. To arrive at the total sample size of 385, the population between 18 and 69 years of age was chosen, with a confidence level of 95%. This means that when we repeat the survey, we would expect the same results in 95% of the cases with a 5% error margin. The sampling was applied in areas at three levels of marginalization in each municipality. The number of questionnaires per municipality was based on the number of inhabitants of each, so that the sample was representative. Zapopan is the municipality with the most inhabitants; it counts 67% of the total inhabitants of both municipalities, so 262 out of 385 questionnaires (representing 67% of the questionnaires) were applied there. Tlaquepaque has only 33% of the inhabitants, thus 123 (33%) questionnaires were applied.

The survey questionnaire was developed to assess the formal environmental education of the participants, the type and quantity of household waste, if and how they separate MSW at the household level, knowledge of contamination, opinions on the MSW collection infrastructure in their city and on the role of citizens in municipal-level MSW management, perceptions of the environmental impacts of MSW disposal, final disposal, environmental impact of MSW, and environmental culture related to MSW. This last one is understood as the cultural practices of the population regarding disposal of their waste, understanding culture as "that complex whole which includes knowledge, belief, art, morals, law, custom, and any other capabilities and habits acquired by man [sic] as a member of society" (Tylor, 1871, p. 1).

Results: consumption, culture, and environmental education

The combined results from the survey, ethnographic observations, key informant interviews, informal conversations, and workshops can be analyzed in terms of two major categories: the consumption patterns and their variation in these municipalities based on levels of marginalization, and the cultures of disposition, including currently lacking environmental education for improving the implementation and future development of MSW disposal policy. Both of these categories are discussed below.

Consumption patterns

In the state of Jalisco, NAE-SEMADES-007/2008 is in force, in which it is mentioned that it is the obligation of every person to separate waste. Although this rule came into effect in 2008, the inhabitants of the municipalities studied do not carry out this activity, and surveys suggest that the lack of separation of MSW may be because residents have not had training in segregation. Further, when the consumption patterns of the MSW and the use of products that cannot be recycled increases, regulation and policies are not applied.

The type of MSW does vary between the communities depending on levels of marginalization; the type of discarded waste in a neighborhood with a high level of marginalization is not the same as that in communities with a low level of marginalization. In communities with a low and medium level of marginalization, apart from the organic and sanitary waste being discarded, there is also a significant amount of PET (polyethylene terephthalate), such as soda bottles, energizers, bottled water, juices, cleaning products, among others, followed by cans of both food and beverages. Likewise, in these communities a high amount of styrofoam and cardboard is generated from food product waste, including pizza boxes and styrofoam food containers. Other types of waste found in these communities were clothing, footwear, accessories, toys, furniture, and household electronic devices.

On the other hand, the waste generated in the communities with a high level of marginalization varied depending on if they were in the urban or rural zone, with overall less plastic or PET waste compared to neighborhoods with low and medium levels of marginalization. These communities also did not throw away garments or footwear, and the waste that they took to the truck was almost unusable. However, in rural communities in the municipality of Zapopan, less organic waste is disposed of than the other studied communities, because residents here use their organic waste as compost or as fertilizer for their farmland. In La Venta, there was more PET in comparison to Huaxtla, because La Venta is located near a principal highway, so people sell a lot of beverages in plastic bottles and the cars that stop in the area throw their bottles away on the streets. Huaxtla is located in the middle of a canyon, and the inhabitants here are concerned about protecting the environment; as a result, the population of this community consumes fewer plastic products.

The generation of MSW as differentiated by social classes is also perceived in landfills and collecting trucks because the scavengers and the collectors know which communities have more recyclable and reusable items that they can sell or use. One garbage collector said:

> If you go to a sanitary landfill, where the garbage collectors know where the garbage comes from, if the truck comes from Las Colinas, no way, it has a lot of valuable items, but if the truck comes from the Cerro del Cuatro, it does not bring anything worth it. Even the amount of waste is also very different.
>
> *(Interview with garbage collector in San Pedro Tlaquepaque, Jalisco, Mexico, in 2016. Translated by the author)*

Garbage collectors know that zones differ in terms of the quality and quantity of MSW and they target areas where more waste is generated that can be sold, reused, or recycled. As collecting recyclable and resaleable waste adds to the income of the garbage collector, this waste is lucrative to them. However, the waste collector cannot choose their own route of waste collection, hence the trade union sends collectors to an area of a lower socio-economic level due to a sanction or punishment, and sends others to an area of a higher socio-economic level or "good zone" as a prize or promotion.

In addition, the waste experts that took part in the Workshop for the Elaboration of the State Program for the Integral Management of Residues of the State of Jalisco identified that the model of consumption in our current society and the lack of community integration in local projects have caused the amount of MSW to increase. However, they also said that society is not entirely at fault for the increase of waste, because the supply of products at a lower price and disposable wrappings by companies cause more MSW pollution. Furthermore, the issues of packaging and wrapping are a sensitive problem within public policy in Mexico because there are no standards or laws that designate the allowed characteristics of the packaging materials of the products consumed in the country. The Colectivo Ecologista de Jalisco (CEJ) worked together with the National Institute of Ecology in 1994 to make a proposal and guidelines for the creation of the Regulatory Law on Packaging derived from Articles 134–143, and other related articles of the General Law of Ecological Equilibrium and Protection to the Environment. However, none of it was ever approved. The Director of the CEJ commented about this event:

> [...] the issue of packaging and wrapping remains a drama. We made a proposal of an official Mexican standard on packaging, wrapping and waste material [...] it was a super avant-garde standard. We got the people of the National Institute of Ecology to help us to promote it. The plastic industry wasn't happy with these standards and made a lot of pressure to block the approval of our standard proposal. [...] At the federal level, you do not have a policy that controls packaging and wrapping. You do not have a policy that internalizes the environmental costs of the disposable [...] when will you really put unpayable taxes on styrofoam?
>
> *(Interview with Director of the CEJ, Guadalajara, Jalisco, Mexico. Translated by the author)*

The culture of waste disposal

Like consumption, the type of municipality does not correspond to differences in the culture of MSW disposal. To some extent, the inhabitants of Zapopan and San Pedro Tlaquepaque are homogeneous in terms of disposal. The differences again are related to the levels of marginalization.

In the rural areas of these municipalities, MSW is burned, which causes a severe environmental impact due to the gases that are emitted. Rural residents are burning not only organic waste, but also PET, paper, styrofoam, and other inorganic materials (see Figure 6.2).

FIGURE 6.2 Burning of waste in La Venta, Zapopan

Likewise, there is still illegal disposal of MSW in abandoned lands or forests, as the Director of the Environment of Zapopan 2015–2018 commented:

> We have a lot of illegal disposal, especially of waste that needs special handling, there are many red points, for example: zone of the Spring, zone of the Colli, of this side of the White River, near the Nixticuil.
>
> *(Interview with the Director of Environment of Zapopan 2015–2018, in 2016. Translated by the author)*

The "zones" or communities indicated by the Director of the Environment of Zapopan correspond to areas of high marginalization and rural communities, where the disposal habits continue to rely on burying, burning, or simple makeshift landfilling.

Another important point in the culture of MSW disposal is that the type of waste people want to dispose of can cause conflicts with collectors. Armchairs, electronics, and other large items do not fit inside the collectors' trucks, and therefore these big items stay behind on the public roads for days. It should be noted that it is not the obligation of the collectors to take such waste as it can damage their vehicle, and in the case of electronics, the waste needs special handling because some items are categorized under hazardous waste, which is not the responsibility of the municipality.

The population also does not respect the collection schedules, which causes delays for the collectors. Since the trash is deposited in plastic bags in the streets, stray dogs destroy the bags and spread the garbage all over the streets. The population is accustomed to disposing of their waste as they please, and this has both environmental and political consequences, bringing problems in the efficiency of the collection, transfer, and final disposal of MSW.

Environmental education refers to the knowledge (acquired formally or informally) that the population has about the environment, and in this particular case about solid waste and its separation, characteristics, and impacts on the environment. Table 6.1 presents results of the survey that addresses the issue of environmental education. As a result of the lack of training received by the population on the classification of waste, only 15% of people separate the garbage in their house, which can be seen in Figure 6.3. As presented in Figure 6.4, 66% of people in the two municipalities throw their food waste with the other garbage, only 18% give it to their pets, and 6% make compost. The importance of this chart is that more than 50% of the total MSW generated is organic material, representing an incredible opportunity to improve disposal to reduce the amount of waste that reaches the landfill.

Another important source of waste that can be minimized is yard waste, including leaves and branches, by making compost. As shown in Figure 6.5, 73% of people bring this type of waste to the collecting truck, although it is not the collectors' obligation to accept this waste. According to the municipal regulation, this kind of waste is from another department (gardens and parks department); this is a clear

TABLE 6.1 Environmental education results

	Yes	*No*	*Don't Know*
In the last five years have you received any training on how to handle the garbage that is generated in your home?	28%	56%	16%
Have you studied any course on waste classification?	33%	52%	14%
Do you know what kind of trash garbage trucks do not take?	54%	-	46%

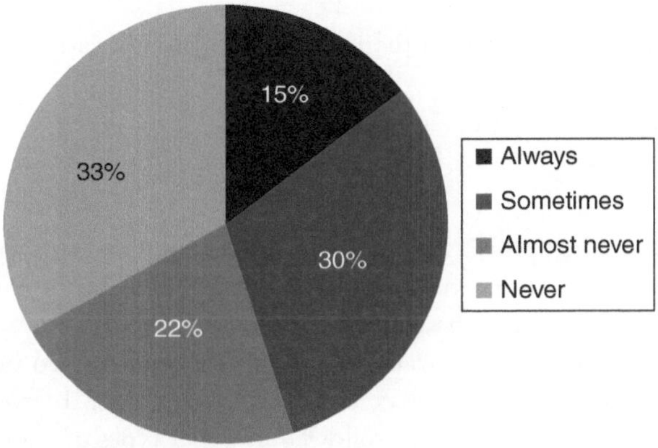

FIGURE 6.3 Do you separate trash in your home?
Source: Beatriz Adriana Venegas Sahagun

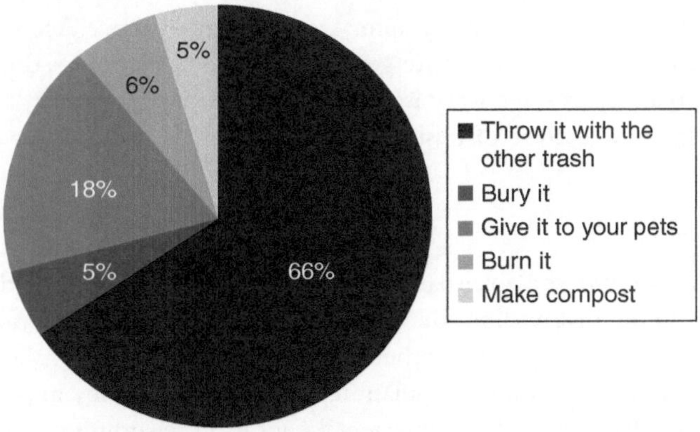

FIGURE 6.4 What do you do with food waste?
Source: Beatriz Adriana Venegas Sahagún

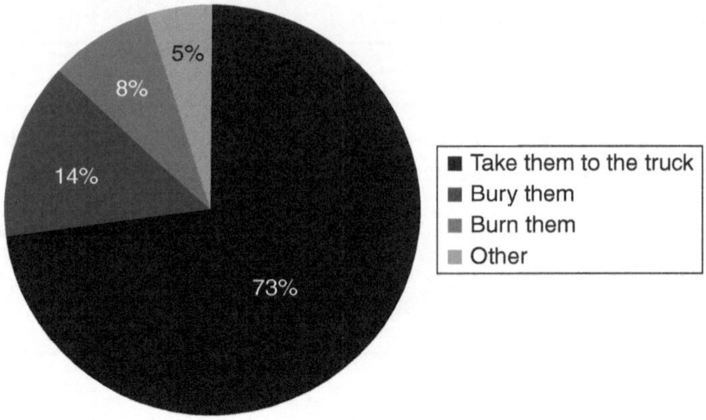

FIGURE 6.5 What do you do with the leaves or branches that are generated at your home?

Source: Beatriz Adriana Venegas Sahagún

example that even when policies are in place, implementation is difficult because of insufficient public support, corruption in the system, and lack of communication between the government and the population.

For 30 years, MSW characterization studies have been carried out in Mexico and in the AMG, where organic waste covers 50% to 58% of the total MSW, followed by paper and cardboard waste, followed by other types of waste like plastics, glass, and metals (Phillips, Restrepo, and Rathje, 1984; Bernache, 2011; Íñiguez-Covarrubias et al., 2011; Secretaría de Medio Ambiente y Recursos Naturales, 2013; Carrasco Escobar and Rodríguez Silva, 2015). Of the waste generated, 28% is potentially recyclable, 19% is not usable, and the rest is organic (Carrasco Escobar and Rodríguez Silva, 2015). These statistics accurately correspond to what was found in the surveys and the ethnographic work presented above. According to the surveys, the composition of the waste generated in the household is dominated by food waste; organic waste recovery is currently a missed opportunity, due to both the structure and the culture of waste disposal.

Conclusions

Citizen participation in the governance of MSW was the main focus of this study; it is important to note that without the population's participation, good management of MSW cannot be done, yet also, if there is no collaboration between all the actors involved including government, population, collectors, etc., policy implementation is deficient. The deficiency of environmental education regarding the responsibilities for separating MSW suggests a lack of collaboration between inhabitants and the city council. Municipalities arguably have not implemented training and awareness strategies because terms of government are very short and incoming municipal managers always have to start from scratch, so they end up only devoting themselves to issues of MSW collecting. This phenomenon of starting from scratch is common in Mexico's local governments because the bureaucracy is not functioning as it should. Each time the administration of the municipalities changes, the files and programs are deleted or destroyed. In other cases, the new administration prefers to start with their own new programs instead of continuing with those of the previous administration. This occurs more often when a different party wins the election. These bureaucratic failures forestall policy continuity and progress.

Consumption habits are an important factor in the generation of MSW, but the government does not directly control consumption, and there are no policies that directly manage household consumption. Also, there is no policy regulating product packaging, which creates a lot of materials that cannot be recycled such as styrofoam, cellophane, and plastic. This is exacerbated by planned obsolescence, which is not addressed in the public agenda and no environmental policies minimize the environmental impact that this generates since the economic factor predominates both production and politics.

Consumption is related to the environmental education that inhabitants have (Jensen and Schnack, 1997). Public policies should be aimed at addressing consumption, conducting public campaigns for separation and composting and regulating packaging. In addition to the above, municipalities should emphasize composting, since organic waste is more than 50% of the total waste that arrives at the disposal site. To carry out composting, municipalities should create awareness programs that promote the benefits of this activity.

Yet in the case of social participation in MSW in the municipalities of Zapopan and San Pedro Tlaquepaque, the population is not aware of the environmental damage caused by not separating, recycling, or reducing their waste. For the population, all that no longer serves is garbage, and residents do not have a culture of disposal in which they seek ways to reduce waste. In areas with a high level of marginalization, waste is valued and utilized, but this is still not a regular activity and priority of the population more generally. The lack of public policy on consumption, packaging, and wrapping, coupled with lacking awareness and environmental education, increases the amount of MSW generated in these communities. Also, the lack of application of existing laws increases the absence of interest from the population, because there are no sanctions. Environmental awareness of the population must be placed on the public agenda of these and other municipalities in Mexico, but for this to happen, society must be empowered, and it must know that it can change the direction of MSW, starting with governance from the bottom of society.

Political power shapes the consumption patterns of a society, and the hidden actors within consumption are found in the large industries that generate MSW, which have agency capacity in decision-making and where "the logic of consumption customary in small communities are intimately linked to broader value regimes defined by large-scale policy" (Appadurai, 1991, p. 48). Consumption is controlled by political power, as evidenced by the absence of laws regulating product packaging even though packaging and wrapping are large contributors to MSW. However, it is also true that policy can work to address structural barriers (like packaging) as well as individual behaviors simultaneously, only by grappling with MSW as a multi-faceted and multi-scalar issue related to social categorizations and social justice with huge implications for the health of the land and human communities.

References

Acedo, B. (2007). La libertad municipal y la democracia. *Nexos*, 29(350), pp. 37–41.

Alonso, L. (2005). *La era del consumo*. Madrid: Siglo XXI.

Appadurai, A. (1991). *La vida social de las cosas*. México: Consejo Nacional para la Cultura y las Artes.

Ayres, R. and Kneese, A. (1969). Production, consumption, and externalities. *The American Economic Review*, 59(3), pp. 282–297.

Baudrillard, J. (1979). *El sistema de los objetos*. México: Siglo Veintiuno Editores.

Belk, R. (2004). The human consequences of consumer culture. In: K. Ekström and H. Brembeck, eds., *Elusive Consumption*. New York: BERG, pp. 67–86.

Bernache, G. (2011). *Cuando la Basura nos alcance. El impacto de la degradación ambiental*. Guadalajara: CIESAS-Occidente, Universidad de Guadalajara.

Bernache, G., Bazdresch, M., Cuellar, J., and Moreno, F. (1998). *Basura y metrópoli. Gestión social y pública de los residuos sólidos municipales en la Zona Metropolitana de Guadalajara, Guadalajara*. Guadalajara: Universidad de Guadalajara.

Bifani, P. (1999). *Medio ambiente y desarrollo sostenible*. 4th ed. Guadalajara: IEPALA Editorial.

Bourdieu, P. (1988). *La distinción*. España: Taurus.

Carrasco Escobar, J. and Rodríguez Silva, J. (2015). Análisis de estadísticas del INEGI sobre residuos sólidos urbanos. *Revista Realidad, Datos y Espacio. Revista Internacional de Estadística y Geografía*, [online] 6(1). Available at: www.inegi.org.mx/RDE/rde_14/rde_14_art2.html [Accessed 8 November 2016].

Carson, R. (1962). *Silent Spring*. Connecticut: Greenwich.

CONAPO, Consejo Nacional de Población (2013). *Índice absoluto de marginación*. México: Consejo Nacional de Población.

Couto, I. and Hernández, A. (2012). Participación y rendimiento de la iniciativa privada en la gestión integral de los residuos sólidos urbanos en la frontera México-Estados Unidos. *Gestión y Política Pública*, 21(1), pp. 215–261.

Dai, Y., Gordon, M., Ye, J., Xu, D., Lin, Z., Robinson, N., Woodard, R., and Robinson, N. (2015). Why doorstepping can increase household waste recycling. *Resources, Conservation and Recycling*, 102, pp. 9–19.

Delgado Huertos, E. (1997). Consumo y medio ambiente. *Tabanque: Revista pedagógica*, 12, pp. 41–66.

Delgado Ramos, G. (2012). Metabolismo social y minería. *Ecología Política*, 43, pp. 16–20.

García Del Castillo, R. (2003). La política de servicios municipales en México: casos y tendencias recientes. In: E. Cabrero Mendoza, coord., *Políticas Públicas Municipales, una agenda en construcción*. México: Miguel Ángel Porrúa and CIDE, pp. 231–264.

Hornik, J., Cherian, J., Madansky, M., and Narayana, C. (1995). Determinants of recycling behavior: A synthesis of research results. *The Journal of Socio-Economics*, 24(1), pp. 105–127.

INEGI (2010). Censo de Población y Vivienda 2010 [online]. Available at: www.inegi.org.mx/est/contenidos/proyectos/ccpv/cpv2010/Default.aspx [Accessed 23 September 2014].

INEGI (2010b). Cartografía Geoestadística Urbana 2010 [online]. Available at: /www3.inegi.org.mx/sistemas/productos/default.aspx?c=265andupc=0ands=estandtg=3604andf=2andcl=0andpf=prodandef=0andct=201100000andpg=2 [Accessed 23 September 2014].

INEGI (2012). Medio Ambiente. Asentamientos y actividades humanas. Residuos sólidos urbanos [Online]. Available at: www3.inegi.org.mx/sistemas/temas/default.aspx?s=estandc=21385 [Accessed 23 November 2014].

Iñiguez-Covarrubias, G., Iñiguez-Franco, F., Martínez-Gutiérrez, G., and Ryckeboer, J. (2011). Separación de residuos domiciliarios para la preparación de compost y su análisis en la producción de pepinos. *Agrociencia*, 45(5), pp. 639–651 [online]. Available at: www.scielo.org.mx/scielo.php?script=sci_arttextandpid=S1405-31952011000500009andlng=esandtlng=es [Accessed 20 November 2016].

Jensen, B. and Schnack, K. (1997). The action competence approach in environmental education. *Environmental Education Research*, 3(2), pp. 163–178.

Miafodzyeva, S. and Brandt, N. (2013). Recycling behaviour among householders: synthesizing determinants via a meta-analysis. *Waste and Biomass Valorization*, 4(2), pp. 221–235.

Miranda Carreno, R. and Blanco Suarez, A. (2010). Environmental awareness and paper recycling. *Cellulose Chemistry and Technology*, 44(10), p. 431.

McCracken, G. (1990). *Culture and Consumption*. USA: Indiana University Press.

Phillips, D., Restrepo, I. and Rathje, W. (1984). El proyecto basura: The archaeology of industrial transformation in Mexico. *American Behavioral Scientist*, 28(1), pp. 139–153.

Rodríguez Becerra, M. and Gabaldón, J. (2002). Evolución de las políticas e instituciones ambientales: ¿Hay motivos para estar satisfechos? In: E. Leff, E. Ezcurra, I. Pisanty, and P. Romero, eds., *La transición hacia el desarrollo sustentable: perspectivas de América Latina y el Caribe*. México: PNUMA, pp. 35–60.

Schultz, P., Oskamp, S., and Mainieri, T. (1995). Who recycles and when? A review of personal and situational factors. *Journal of Environmental Psychology*, 15(2), pp. 105–121.

Secretaría de Medio Ambiente y Recursos Naturales (2013). *Informe de la situación del Medio Ambiente en México*. México: SEMARNAT.

Stillerman, J. (2015) *The Sociology of Consumption: A Global Approach*. Cambridge: John Wiley and Sons.

Timlett, R. and Williams, I. (2011). The ISB model (infrastructure, service, behaviour): A tool for waste practitioners. *Waste Management*, 31(6), pp. 1381–1392.

Tylor, E. (1871). *Primitive Culture*. New York: Harper.

PART III

Human health and well being

7

WHERE IS THE JUSTICE?

An examination of the failure of the US EPA Office of Civil Rights to ensure environmental justice for poor and minority communities

Amelie Simons

In 1993, the Clinton administration established the Environmental Protection Agency's (EPA) Office of Civil Rights (OCR) as the principal agency ensured with the charge of equal treatment and application of environmental regulations by EPA fund recipients for all United States communities. Yet over the course of more than two decades and 300 complaints of environmental harm in poor and minority communities, the OCR has yet to make a single formal finding of environmental discrimination according to Title VI Civil Rights law that provides a statutory basis for seeking relief from any discrimination a community or individual may face from federally funded programs or activities (USCCR Report, 2016; USDOJ, 2015; USDOJ, 2016). The fact that the OCR has neglected to recognize any instance of environmental discrimination despite myriad cases that suggest otherwise indicates the possibility of not only an agency operations issue but also the existence of a severe impediment to implementing environmentally just policies in the United States.

The preliminary evidence of the OCR failure to recognize environmental discrimination originated from a 2016 review by the United States Commission on Civil Rights (USCCR) concerning the role of the EPA in environmental justice titled "Environmental Justice: Examining the Environmental Protection Agency's Compliance and Enforcement of Title VI and Executive Order 12,898" (USCCR Report, 2016). The goal of this chapter is to provide a comprehensive understanding of the failure of the OCR to recognize and remediate environmental discrimination, beginning with an introduction to the topic of environmental justice followed by a brief overview of the history of environmental justice in the United States. Next, the organizational structure and the objectives of the OCR are outlined from a public management perspective. A case study of the water contamination crisis in Flint, Michigan, is presented as evidence of environmental racism in the United States.

Analyzing this crisis from a public management perspective, this chapter identifies three possible explanations of the OCR failure to ensure environmental justice: (1) the burden of proof for discrimination in the statutory law may simply be too high; (2) there may be a problematic agency culture within the OCR; or (3) the OCR failure may be a symptom of institutionalized racism that is perpetuated through administrative evil. The chapter explores each of these possible explanations and concludes by advocating for the elevation of environmental justice as an issue of primary importance to the EPA. Recommendations for solutions to the issue include a targeted reform of the burden of proof in the statutory law, a reform of the OCR (and larger EPA) organizational culture, and further investigation into the possible presence of administrative evil and institutionalized racism within the agency.

Background

In the early 1960s, Rachel Carson published *Silent Spring*, a book that garnered significant attention and has been credited as the initial inspiration for the environmental movement in the United States (EPA, 2016c). In response to the public demand for more stringent environmental regulation resulting from the increasingly public displays of environmental degradation and the bourgeoning environmental movement in the United States, the Nixon administration established the EPA in 1970 as a means to consolidate environmental protection under the umbrella of a single federal agency (EPA, 2016c; 2016e). The agency's responsibilities largely center on environmental research, the setting of standards and regulations, and the enforcement of environmental protection (USCCR Report, 2016).

Beginning in the mid-1980s, at the same time as critical federal pollution regulations like the Clean Air Act (1970) were being implemented, environmentalists and the public alike began to notice a disturbing pattern, now termed "environmental racism," where incidences of pollution and environmental harm clustered around poor and minority communities (Rhodes, 2003). Though no definitive definition of environmental justice exists, and the terms "environmental justice," "environmental discrimination," and "environmental racism" are often used interchangeably, sociologist Robert Bullard, a central figure in environmental justice literature, defines environmental injustice as "[A]ny policy, practice, or directive that, intentionally or unintentionally, differently impacts or disadvantages individuals, groups, or communities based on race or color; [as well as] exclusionary and restrictive practices that limit participation by people of color in decision-making boards, commissions, and staffs" (USCCR Report, 2016). Rhodes (2003, p. 6), who prefers the term "environmental racism," frames the concept in terms of decision-maker neglect, as well as racially motivated decision-making in both the public and private sectors:

> In this "environmental racism", persons of color, either by design or neglect, endure a disproportionate share of the cost of economic development and growth, without enjoying a corresponding share of either the economic or environmental benefits. They have virtually no representation among the

major nongovernmental environmental organizations and below-average representation within environmental agencies such as the US Environmental Protection Agency.

Regardless of the exact definition used to describe environmental justice, discrimination, or racism, three key characteristics are evident. First, environmental discrimination constitutes the unequal application of government action and/or policy related to the environment. Secondly, such unequal application will follow a pattern that demonstrates a disproportionate neglect or burden for poor and minority communities. Finally, the communities in question will be effectively disempowered and disenfranchised through the lack of representation or other impediments to civic participation. The final characteristic can create conditions of government inaction resulting from low representation completing a vicious cycle.

Although environmental discrimination certainly existed prior to the public recognition of this problem in the late twentieth century, its formal recognition by the federal government did not take place until the 1990s (Rhodes, 2003). Originally, the EPA was intended to be a pollution control agency, not a monitoring body for environmental justice issues (USCCR Report, 2016). However, local-level, grassroots community advocacy in the 1980s helped to vault the problem of environmental racism into the public spotlight and onto the agenda of federal policy makers (USCCR Report, 2016). As a result, the Clinton administration established the EPA's OCR as the principal agency responsible for environmental justice in the United States in 1993 (EPA, 2016a; Newkirk II, 2016), enforcing federal civil rights laws that prohibit discrimination. The OCR is tasked with ensuring equal treatment in the development, implementation, and enforcement of environmental law, policy, and regulation (EPA, 2016b; Newkirk II, 2016). In 1994, President Clinton issued a Presidential Executive Order (number 12,898) requiring federal agencies to incorporate the principles of environmental justice into their objectives as well as to collect data on the impact of federal programs and activities on the human and environmental health of poor and minority communities (EPA, 2016f; USCCR Report, 2016).

The formal complaint system for instances of environmental discrimination exists under the umbrella of Title VI of the 1964 Civil Rights Act (EPA, 2016d). Title VI forms the statutory basis for individuals or groups seeking relief from discriminatory activity by recipients of federal funding or through the activities of the federal government itself (USCCR, 2016). Title VI defines discrimination, creates a complaint mechanism for cases of federally funded discrimination, and defines the burden of proof for the formal recognition of instances of environmental justice. However, the 2016 USCCR report was conducted with the purpose of providing an external review to determine whether the OCR was complying with its environmental justice obligations and discovered that, while over 300 complaints of disproportionate environmental harm in poor and minority communities have been brought to the OCR, the agency has yet to make a single formal finding of Title VI discrimination (Newkirk II, 2016). The failure of the OCR to recognize any instance of environmental discrimination, despite hundreds of cases that claim otherwise, suggests not only that the OCR and the EPA have been ineffective in acknowledging and remediating environmental injustices, but also that a serious obstacle to environmental justice exists in the United States. Existing policy tools intended to enable monitoring of and compliance with environmental justice do not seem to be functioning as intended. The remaining part of this chapter explores the dimensions of public management that may be working to hinder the ability of OCR to fulfill its mandate to address environmental injustices and what other policy-driven entities can learn from these lessons to improve the social justice implications of environmental policymaking.

The EPA OCR and complaints process

The EPA OCR (formerly the Office of Equal Opportunity) is the primary agency responsible for reviewing and investigating Title VI complaints and enforcement issues related to environmental justice, and reviews claims of environmental discrimination under the External Compliance Program. The OCR is located within the Office of the Administrator as part of the larger structure of the EPA. The OCR is responsible for overseeing the EPA's equal employment programs in addition to providing policy and technical assistance to the agency's main office, regional offices, and research laboratories, ensuring that EPA offices around the country comply with Equal Opportunity Employment laws. The primary objective of the program is to ensure that all EPA fund recipients comply with Title VI requirements (EPA, 2016a; USCCR Report, 2016).

Any person or organization within the United States who believes that environmental discrimination is being perpetuated by an agency that receives federal funding from the EPA is eligible to file a complaint with the OCR. The complainant need not be the victim of discrimination (USCCR Report, 2016). In order to file a complaint with the OCR External Compliance Program, four basic requirements must be fulfilled: (1) the complaint must be completed in writing; (2) it must identify the agency that is committing the alleged discrimination; (3) it must identify discrimination according to the parameters of the laws that the OCR oversees (on the basis of race, color, national origin, sex, disability, or age); and (4) the complaint must be filed within 180 days of the date of the last alleged act of discrimination. Once a complaint has been filed, the organization in question has the right to request a hearing to determine whether or not the organization complies with Title VI. In almost all cases of alleged environmental discrimination, parties involved are offered the option to pursue alternative dispute resolution in the form of mediation. Only when all parties agree to mediation is mediation adopted as a course of action.

Numerous instances of alleged environmental racism exist in the United States, as do shocking statistics concerning the disparate application and enforcement of environmental regulation among polluters operating in predominantly minority communities. Rhodes (2003, p. 5) provides examples of some of the most disconcerting statistics from the era of the inception of the OCR to evidence the existence of environmental racism:

> Fines imposed on polluters by all levels of government in white communities in the 1980s were 46 percent higher than those imposed for violations in minority communities... Between 1990 and 1993, of more than 22 community grant requests to study the feasibility of locating a nuclear waste facility within a locale, 16 came from Indian reservations.

The figures cited represent only a small portion of the evidence that supports claims of systematic environmental racism in the United States. Numerous case studies exist across the country, which suggests the existence of a pattern of environmental racism, particularly in the selection of localities for hazardous waste disposal. In recent times, the lead crisis in Flint, Michigan, has been identified as one major example of environmental injustice in a minority community consisting of majority black or African-American population living in one of the most economically impoverished communities in any US metropolitan areas (FWATF, 2016). Using the Flint water contamination crisis as a case study, the remaining part of the chapter will focus on the failure of OCR to handle cases of environmental injustices.

Case study of OCR handling of environmental racism: Flint, Michigan

Likely the most notable instance of purported environmental racism in recent time is the Flint, Michigan, water contamination crisis. Flint, a small city of around 100,000 predominantly black or African-American residents, is situated in eastern Michigan (US Census, 2015). In April 2014, the city of Flint decided to switch the primary water supply source from the shared system out of Detroit (Lake Huron) to the Flint River as a cost-saving strategy in the economically depressed community. In February 2015, water samples indicated a high level of lead contamination of around 104 parts per billion (ppb), significantly higher than the EPA testing levels of 15ppb (Hannah-Attisha et al., 2016). Though the Michigan Department of Environmental Quality (MDEQ), a EPA fund recipient organization, was in charge to enforce federal laws, the EPA delayed the enforcement of the Safe Drinking Water Act and Lead and Copper Rule, despite some shockingly high findings of lead contamination in the city water of up to almost 400 ppb, being hesitant to draw any conclusions or recommend that residents use alternative sources of water (Dana and Tuerkheimer, 2016; Lin et al., 2016).

In September 2015, a Virginia Tech researcher posited that the Flint River's corrosive water was breaking down the antiquated water distribution lead pipe system, leeching lead into the water supplies (Dana and Tuerkheimer, 2016). However, the MDEQ disputed these findings as unsubstantiated. It was not until late 2015 that a third-party researcher became the first scientific authority to urge residents to stop using water supplies from the Flint River. Following the recommendation,

Flint reconnected to the Detroit system. However, this switch did not completely remedy the contamination, and in December 2015, the city of Flint declared a state of emergency. In January 2016, President Obama declared a federal state of emergency in Flint, and federal funding was allocated to the city through the Federal Emergency Management Agency for aid (Lin et al., 2016).

Throughout the course of events, evidence came to light that government agencies, particularly the MDEQ, an EPA fund recipient, were primarily to blame for the crisis. The MDEQ not only failed to implement proper corrosion control but also facilitated the long duration of the contamination crisis with their hesitancy to issue an early recommendation that residents use alternative sources of water (Lin et al., 2016). Throughout the course of the crisis, the residents experienced the immediate health consequences of lead poisoning, including rashes and breathing problems (Christensen and Sidner, 2016). More recently, those affected have also begun to experience some of the severe long-term and irreversible health consequences of lead poisoning. Additionally, numerous children that consumed city water during the contamination crisis now carry the potential long-term risk of negated brain and nervous system development (Christensen and Sidner, 2016).

As of July 2016, the EPA had yet to initiate a formal investigation into the potential civil rights violations of the Flint water crisis despite the fact that the majority of victims are black or African-American, a racial minority in the United States (Leven, 2016). However, at the time of this writing, the agency insists that it is reviewing complaints of Title VI environmental discrimination to determine whether or not it can proceed in a formal civil rights investigation under Title VI statutory law. Residents of Flint have filed complaints with the OCR, which their attorneys insist should easily fit the four requirements for consideration (Leven, 2016). Nevertheless, critics of the EPA have expressed doubts about the potential for a formal finding of discrimination, as one such critic states:

> To make a finding of disparate impacts in Flint, the EPA would need to compare the approach in Flint to that taken in non-minority communities in the state, and then try to line up the facts as close as possible to show difference between actions in different communities... People also have raised Flint-like scenarios to the EPA in the past in complaints and the EPA has found ways not to act.
>
> *Leven, 2016*

Reservations like this about the dubious possibility of a formal finding of environmental injustice suggest that the burden of proof for environmental racism is set impossibly high, given that there is unlikely to be a perfect case for comparison to demonstrate different and discriminatory choices leading to different impacts. However, this is only one possible explanation for the lack of success in litigating environmental injustice through the OCR, and other possible limitations to the effectiveness of social justice via environmental policy also deserve exploration.

Analysis of the issue

There are at least three potential explanations for the failure of OCR to make a single finding of Title VI discrimination, even in cases where the evidence of devastating impacts on vulnerable populations is as stark as it is in the case of Flint, Michigan. The first option is that the standard of proof for Title VI, Section 601 adjudication is simply set too high to allow any complaint to meet the requirements (Newkirk II, 2016). This is what is implied in the quote presented above. The second option posits that, since its origin, the OCR has experienced such agency dysfunction in the organizational culture that it has completely failed in its primary objective. The final possibility is the most nefarious and suggests some sort of administrative evil that has perpetuated the systematic inequalities that underlie the history of institutionalized racism in the United States (Griffith et al., 2007, p. 287).

The OCR set a precedent early in its history to issue no formal findings of discrimination in response to Title VI complaints. For example, between 1993 and 1998, 58 Title VI complaints were filed, with the EPA failing to uphold a single complaint (USCCR Report, 2016). According to a 2003 OCR document "Not in My Backyard," the reason for the early-set precedent of no formal findings was that in most instances during that period, either the OCR failed to properly investigate the complaints, or the complaints were dismissed for jurisdictional or technical reasons (USCCR Report, 2003, p. 31). This failure may be primarily the result of too high a burden of proof in the Title VI, Section 601 statutory law, which requires that the discrimination must be demonstrated to be intentional (USCCR Report, 2016).

Furthermore, the scope of Title VI protection has been another cause for criticism. The law provides protection from discrimination on the grounds of race, color, or national origin (Title VI of the 1964 Civil Rights Act, Section 601). According to the USCCR Report (2016), "because Title VI only addresses discrimination based on 'race, color, or national origin', the statute offers no remedy to communities which cannot make a prima [facie] of disproportionate demographic make-up… Title VI remedies are apparently unavailable to marginal communities which cannot show that racial or ethnic discrimination drove the actions of a Title VI recipient in choosing or approving the site" (p. 11). In contrast to the types of discrimination covered by the law, market-based evidence exists to suggest that environmental discrimination does not only follow a pattern in racial and ethnic minority communities, but also communities that are vulnerable according to socio-economic factors. Supporters of what is known as the market forces argument insist that poor communities tend to accumulate around polluting facilities due to low property values in these areas (USCCR Report, 2016). Often, racial and socio-economic forms of marginalization intersect in communities affected by environmental discrimination. Thus, the USCCR recommends that racial, social, and economic structures be viewed as intersectional components rather than independent concerns (USCCR Report, 2016). However, this is not the case under

Title VI protections, as the racial and ethnic demographic of a community is considered while other socio-economic factors are not.

In addition to Title VI, environmental justice is also guaranteed as part of Presidential Executive Order 12,898. As a formal recognition of the importance of environmental justice, this 1994 executive order from President Clinton was intended as a measure to help focus federal attention on the environmental and human health outcomes of federal actions in poor and minority communities (EPA, 2016f). By requiring federal agencies to collect data on the health and environmental outcomes of their actions, the executive order attempted to force the incorporation of environmental justice considerations into federal permitting decisions (USCCR Report, 2016). Moreover, with the executive order, Clinton intended to bring the issue of environmental justice to the attention of federal agencies and incorporate consideration for this issue into agency practices (EPA, 2016f). On this front, the executive order was largely successful. However, the measure does not remedy the issue of the Title VI protection gap for low income and marginalized communities, as it does not create legally enforceable substantive or procedural rights (USCCR Report, 2016). Instead, the executive order functions as symbolic and aspirational policy that recognizes the need for federal agency awareness of environmental justice issues. As such, the gap in the protection for vulnerable communities that do not meet the Title VI requirements still exists today.

The possibility of an unworkable and impossibly high burden of proof within statutory law constitutes a violation of accountability between public institutions, public officials, and the public constituency. According to Hill and Lynn (2016), the rule of law is the foundation of accountability between the government and the people. If the burden of proof in Title VI, Section 601 is set to prevent the workable execution of the rule of law regarding environmental justice, then a serious impediment exists to the assurance of accountability between the government and the public. The lack of accountability on this front compromises the integrity, performance, and legitimacy of the government to fulfill its obligations to protect and ensure environmental justice. Compromised accountability also diminishes the overall capacity of the federal government to maintain democratic control over environmental justice issues. Finally, decreased accountability prevents catharsis and thus the healing of the compromised relationship between the government and victims of environmental racism. The obstacle to catharsis is especially apparent in light of a malfunctioning complaints process, which is supposed to ensure justice for victims of environmental discrimination.

Another potential explanation for the failure of the OCR to make a formal Title VI finding of discrimination lies in the organizational culture of the EPA. Hill and Lynn (2016) claim that the EPA has a strong tradition and thus an established organizational culture of caution, particularly in policy development. The possibility that the EPA organizational culture and management structure is at least partially to blame for the lack of formal Title VI violations is corroborated by the findings presented in the USCCR Report (2016), which found that the EPA has a history of failure to meet deadlines, extreme delays in responding to Title VI complaints, and a lack of clarity in its mission and a lack of oversight and accountability between the EPA main office and smaller EPA entities, like the OCR.

The problematic organizational culture in the EPA may have perpetuated disconnection between subordinate entities, lack of mission clarity, and a lack of timeliness in responding to Title VI complaints. As such, the Commission argues that the EPA has not substantively incorporated environmental justice as a right into its procedures and decision-making process (USCCR Report, 2016). The EPA's small staff and lack of resources exacerbate this issue (USCCR Report, 2016). Whether these troublesome trends are the direct result of a problematic organizational culture within the EPA has not yet been conclusively determined; however, there is evidence that this is likely the case. In a 2015 blog post, Stan Meiburg, the Acting Deputy Administrator for the EPA in 2015, acknowledged the existence of internal agency impediments to the realization of environmental justice (Meiburg, 2015). In this post, Meiburg proposes a rule change as the remedy for the problems associated with the Title VI complaint system. The suggested rule change involves removal of inflexible, non-statutory deadlines from the internal procedures, which would in turn promote a systematic change in the way the OCR deals with complaints (Meiburg, 2015). Additionally, Meiburg promises that the OCR will be more proactive in its future compliance efforts. While it remains to be seen whether or not such promises will be implemented, it seems unlikely that such superficial changes will be sufficient to combat a potentially problematic and deeply rooted EPA culture. In such cases, public administration literature recommends more drastic measures such as widespread reorganization or the implementation of agency-level rule making as solutions to reform organizational culture (Hill and Lynn, 2016).

The final and likely the most difficult possibility to definitively pinpoint as the source of OCR failure is the existence of a pattern of institutionalized racism in the EPA as well as throughout the federal government, which has been perpetuated through acts of administrative evil. Institutional racism is a systematic set of patterns, procedures, policies, and practices that exist within legitimate institutions and consistently disadvantage minority groups (Griffith et al., 2007). According to perspectives emphasizing institutional racism, the United States maintains a deep, structural racism that has not been altered by the implementation of civil rights laws (Griffith et al., 2007). Consequently, government institutions like the EPA can actually be tools of oppression, as they unintentionally function to strengthen the marginalization and vulnerability of the same minority communities that they have pledged to serve in their agency mission statements (Griffith et al., 2007). This type of institutionalized oppression, whether intentional or not, represents an "administrative evil," which is a social phenomenon that allows legitimate institutions to unconsciously perpetuate the oppression of targeted groups (Adams and Balfour, 2015).

One of the ways in which administrative evil is perpetuated in a government institution is through the norm of technical rationality (Adams and Balfour, 2015). The EPA is an agency in which technical rationality is the predominant culture. The ubiquity of technical rationality within the EPA is especially evident within agency operations concerning toxic waste disposal, an EPA duty that Rhodes (2003) had identified as rampant with instances of environmental racism. It is not impossible to surmise that technical rationality is administrative evil in the form of institutionalized environmental racism.

Moreover, it is possible that this "evil" has survived unnoticed for the majority of the agency's history because minority communities are persistently excluded from the environmental policymaking and decision-making processes. Minority groups have traditionally been under-represented in environmental government and non-governmental institutions and in environmental policymaking (Rhodes, 2003). However, it is incredibly difficult to identify or prove administrative evil (Adams and Balfour, 2015). Thus, while it is conceivable that this phenomenon could play a role in the failure of the OCR to formally identify any case of environmental racism, the possibility that administrative evil will be decisively proven as the impetus for this failure is unlikely.

Conclusion

Clearly, further research is needed in order to substantiate a claim that statutory law, agency culture, or administrative evil constitutes the primary reason why the OCR has never issued a formal finding of Title VI environmental discrimination. However, the evidence presented does suggest that all three explanations provide insight and that specifically the statutory law as well as a problematic EPA culture constitute significant empirical barriers to the issuance of formal Title VI violations.

Further research in the area of environmental discrimination in the United States, and more specifically into the efficacy of the OCR as the primary agency responsible for overseeing and enforcing environmental justice, is needed. While problematic culture, the perpetuation of institutionalized racism through administrative evil, or a combination of the two are possible explanations for the failure of the OCR to protect and ensure environmental justice, more research would be needed to definitively demonstrate that this is the case.

Regardless of the reasons for the lack of formal findings of Title VI violations, it is clear that the OCR is not functioning to ensure and enforce environmental justice in the United States. As is evidenced in the Flint, Michigan, case study as well as numerous other instances of disproportionate environmental harms in poor and minority communities (Bullard and Lewis, 1996; Fothergill et al., 1999; Szasz, 1994; USCCR Report, 2016), environmental racism is a serious problem that is deserving of public and institutional attention. Moreover, it is an issue that is in desperate need of a solution, as severe human health consequences, as was the case in Flint, form a common result of unchecked environmental injustice. The evidenced inefficacy of the OCR as the primary agency in charge of addressing claims of environmental discrimination means that environmental justice is far from guaranteed in the United States. Improved civil rights governance is necessary because, despite the existing laws, environmental standards are clearly not being upheld at the same level for everyone (USCCR Report, 2016).

Nevertheless, the actualization of environmental justice enforcement through Title VI has the potential to be an important tool. Environmental racism is a critical issue deserving of attention and some level of reform is necessary to improve the current system. Without some kind of reform, the health of vulnerable human populations and ecosystems as well as the legitimacy of environmental justice governance are at stake. Moreover, if the origins of the OCR failure to ensure environmental justice do in fact rest in the existence of administrative evil, then there exists a much larger problem of institutionalized environmental racism in the United States.

Even though administrative evil is difficult to prove, race-based inequalities that continue to exist in the United States, and perhaps in many other parts of the world, infiltrating even into organizations that were committed to undoing it, pose serious challenges for environmental policies in pursuit of sustainability. Sustainability is not possible if class-based inequalities and marginalization continue to push out sections of the population into the peripheries of social, economic, and environmental systems and a balance is achieved in the core. Therefore, social equity and social justice is a prerequisite for sustainability. However, as we see from the case of Flint and in other discussions in the chapter, social inequality and cases of injustices cannot be addressed as long as representation does not increase from minority communities in decision-making platforms. Policy makers are often flooded with causes that need decisions, and only concerted effort from policy actors get some issues to fruition into policies. Moreover, even when policies materialize, many are symbolic in nature, laying down aspirational goals that the policy aims to attain, leaving the implementation plan to be developed by subsequent implementing agencies. As a result, the policy goals can get interpreted differently from their original intent and purpose, at times producing less than expected or opposite results as we see in the case of OCR. Therefore, to pursue sustainability, our socio-political regimes also need structural changes.

References

Adams, G. and Balfour, D. (2015). *Unmasking Administrative Evil*. 4th ed. Armonk, NY: M.E. Sharpe.

Bullard, R. D. and Lewis, J. (1996). *Environmental Justice and Communities of Color*. San Francisco: Sierra Club Books.

Christensen, J. and Sidner, S. (2016). Flint and lead poisoning: Living with it and uncertainty, long-term. CNN.com [online]. Available at: www.cnn.com/2016/03/04/health/flint-lead-exposure-long-term-pain/ [Accessed 9 August 2017].

Dana, D.A. and Tuerkheimer, D. (2016). After Flint: environmental justice as equal protection. *Northwestern University Law Review*, 111(3), p. 879.

Fothergill, A., Maestas, E. G., and Darlington, J. D. (1999). Race, ethnicity and disasters in the United States: A review of the literature. *Disasters*, 23(2), pp. 156–173.

FWATF (2016). Flint Water Advisory Task Force Final Report. Michigan.gov [online]. Available at: www.michigan.gov/documents/snyder/FWATF_FINAL_REPORT_21March2016_517805_7.pdf [Accessed 9 August 2017].

Griffith, D., Childs, E., Eng, E., and Jeffries, V. (2007). Racism in organizations: The case of a county public health department. *Journal of Community Psychology*, 35(3), pp. 287–302.

Hanna-Attisha, M., LaChance, J., Sadler, R. C., and Champney Schnepp, A. (2016). Elevated blood lead levels in children associated with the Flint drinking water crisis: a spatial analysis of risk and public health response. *American Journal of Public Health*, 106(2), pp. 283–290.

Hill, C. and Lynn Jr. L. (2016). *Public Management: Thinking and Acting in Three Dimensions*. 2nd ed. Los Angeles: Sage.

Leven, R. (2016). Flint crisis warrants EPA civil rights review: attorneys. Bloomberg BNA [online]. Accessed at: www.bna.com/flint-crisis-warrants-n73014444081/ [Accessed 9 August 2017].

Lin, J. C. F., Rutter, J., and Park, H. (2016). Events that lead to Flint's water crisis. *New York Times* [online]. Accessed at: www.nytimes.com/interactive/2016/01/21/us/flint-lead-water-timeline.html [Accessed 9 August 2017].

Meiburg, S. (2015). Announcing a series of actions to strengthen EPA's civil rights program. EPA Connect [online]. Accessed at: blog.epa.gov/blog/2015/12/announcing-a-series-of-actions-to-strengthen-epas-civil-rights-program/ [Accessed 9 August 2017].

Newkirk II, V. (2016). The EPA's failure to protect people from the environment. Government Executive [online]. Accessed at: www.govexec.com/management/2016/09/epas-failure-protect-people-environment/132013/ [Accessed 9 August 2017].

Rhodes, E. (2003). *Environmental Justice in America: A New Paradigm*. Bloomington, IN: Indiana University Press.

Szasz, A. (1994). *Ecopopulism: Toxic Waste and the Movement for Environmental Justice*. Minneapolis, MN: University of Minnesota Press.

US Census Bureau (2015). Quickfacts: Flint city, Michigan [online]. Accessed at: www.census.gov/quickfacts/table/PST045215/2629000 [Accessed 9 August 2017].

US Commission on Civil Rights (USCCR) (2003). Not in My Backyard: Executive Order 12,898 and Title VI as Tools for Achieving Environmental Justice. Accessed at: www.usccr.gov/pubs/envjust/ej0104.pdf [Accessed 9 August 2017].

US Commission on Civil Rights (USCCR) (2016). Environmental Justice: Examining the Environmental Protection Agency's Compliance and Enforcement of Title VI and Executive Order 12,898. Accessed at: www.usccr.gov/pubs/Statutory_Enforcement_Report2016.pdf [Accessed 9 August 2017].

US Department of Justice (2015). Title VI of the 1964 Civil Rights Act. Accessed at: www.justice.gov/crt/title-vi-1964-civil-rights-act [Accessed 9 August 2017].

US Department of Justice (2016). Title VI of the Civil Rights Act of 1964 42 U.S.C. § 2000D ET SEQ. Accessed at: www.justice.gov/crt/fcs/TitleVI-Overview [Accessed 9 August 2017].

US Environmental Protection Agency (EPA) (2016a). About the Office of Civil Rights (OCR). Accessed at: www.epa.gov/aboutepa/about-office-civil-rights-ocr [Accessed 9 August 2017].

US Environmental Protection Agency (EPA) (2016b). Alternative Dispute Resolution (ADR) by EPA's Administrative Law Judges. Accessed at: www.epa.gov/alj/adr [Accessed 9 August 2017].

US Environmental Protection Agency (EPA) (2016c). EPA History. Accessed at: www.epa.gov/history [Accessed 9 August 2017].

US Environmental Protection Agency (EPA) (2016d). How to File a Complaint of Discrimination brochure. Accessed at: www.epa.gov/ocr/how-file-complaint-discrimination-brochure [Accessed 9 August 2017].

US Environmental Protection Agency (EPA) (2016e). Milestones in EPA and Environmental History. Accessed at: www.epa.gov/history [Accessed 9 August 2017].

US Environmental Protection Agency (EPA) (2016f). Summary of Executive Order 12,898 – Federal Actions to Address Environmental Justice in Minority Populations and Low-Income Populations. Accessed at: www.epa.gov/laws-regulations/summary-executive-order-12898-federal-actions-address-environmental-justice [Accessed 9 August 2017].

8

THE POWER OF THE TALKING POINTS

Persuasive power and the challenges of sustainable natural resource development

Amanda Kreuze, Roman Sidortsov, and Chelsea Schelly

Between 2000 and 2016, shale gas production in the United States increased from approximately 2 billion cubic feet of daily production to approximately 42 billion cubic feet (US EIA, 2016). Much of this development has occurred in the Midwestern United States, where approximately half of US shale production currently takes places (US EIA, 2016; Ridlington et al., 2016). Traveling through some of the rural parts of this region may elicit two common sights: wellheads and roadside bars. This chapter is inspired by a particular event in which these two common landmarks intersected. When conducting research on community perceptions of local development of high-volume hydraulic fracturing activities, a participant confidentially shared a hard copy of "Talking Points for Selling Oil and Gas Lease Rights" (the "Guidelines"), several pages of typed notes marked "Proprietary," clearly intended to guide landmen in the process of talking to landowners about selling access to the resources that potentially lie under their feet (Kreuze et al., 2016).

Companies aggressively obtain oil and gas leases throughout the Midwest and the Guidelines may play an influential role in making them successful. Estimating their influence is difficult, as not every state requires companies to record their leases. However, the Guidelines were originally provided to us in Michigan, reference Ohio, and were found verbatim elsewhere and later posted online (Archive, n.d.); thus, we can imagine their influence is pervasive across the Midwestern US. In contemplating the role of these notes in the countless conversations they have likely guided, existing theories of power are inadequate for fully understanding how they embody inequitable social power and the implications of that for sustainable, healthy communities.

Attentiveness to the role of power relations in shaping choices about the use of natural resources is necessary. Power is an inevitably tricky concept, ubiquitous yet often ambiguous, contested and multi-dimensional (Lukes, 2005). Various theories of power have dominated scholarly understandings at different times, and many different views abound (Lukes, 2005). Our goal is to illustrate a conception of power based on persuasion, or persuasive power, and to theorize its role in shaping the potential for sustainable, healthy communities. The terminology and idea surrounding the power of persuasion is not new; it is often applied when understanding the role of institutions such as the media (Gamson et al., 1992; Ladd and Lenz, 2009). Others have written specifically about power as legitimation (Freudenburg and Alario, 2007).

We argue that persuasive power as an analytical tool sheds new light on the challenges to protecting community health and well being. Persuasive power often involves legitimation through implied connection to sovereign power, which can be employed to shape community and individual decision-making even in the absence of a legal foundation. Persuasive power persuades via assumed connection to sovereign institutions, law for example, even when it may be working in contradiction to the very structures of power deemed legitimate by communities that imagine them to be working for a public good. It operates by employing sovereign power as a form of discursive framing to legitimate persuasion. By discursively setting the limits of conversation and by implying a link to the sovereign, sustainable natural resource management is limited by the use of persuasive power to shape outcomes. Additionally, these processes present challenges for communities seeking to improve human health and well being via equitable distribution of environmental resources.

Conceptualizing power and law

According to Bertrand Russell (1938), power is "the fundamental concept in social science," comparable in its importance to that of the concept of energy in physics (p. 10); it is arguably impossible to understand the social world without contending with power. The scholarship on power is vast and diverse, with many scholars offering divergent views on what constitutes power in the social context (Morris, 2002). Manuel Castells (2011) defines power as "the relational capacity of a social actor to influence asymmetrically the decisions of other social actors in ways that favour that power actor's will, interests, and values" (p. 10). John Scott (2001, p. 16) explains that the persuasive influence of power works through "cognitive symbols" and through "value commitments" and differentiates between signification (winning the minds) and legitimation (winning the hearts) as the two principal forms of persuasion. Therefore, a persuasion-premised power relationship should be premised on the freedom of choice and, therefore, should feature plentiful capacity for the principal and the subaltern to wield and resist (Scott, 2001). However, a power relation that is *de jure* based on persuasive influence can *de facto* be perceived by a subaltern as premised on a corrective action (coercion and manipulation) by a sovereign. In the case of natural resource development, persuasive power often operates through structurally defined inequity in power relations, where the use of rhetorical forms amplifies inequitable structural relations while legitimizing a principal through implied connections to sovereign power.

Similar to power, law is not an easy concept to define. *Law* refers to a rule or a set of rules promulgated by a sovereign entity. In *Economy and Society*, Max Weber puts law into the category of sovereign power or as he refers to it "domination by virtue of authority" (Grewal, 2008, p. 117). Notably, Weber includes non-state organizations like corporate entities and churches in the scope of entities capable of exercising sovereign power (Grewal, 2008). Pierre Bourdieu (1987) builds on the Weberian idea of "formal rationality" by recognizing law's "power of form" (p. 809). Bourdieu's inclusiveness comes from his visualization of law as the "juridical field" that he sees as "an area of structured, socially patterned activity or 'practice', in this case disciplinarily and professionally defined" (Bourdieu, 1987, p. 806). Thus, law delimits the boundaries of social power and counter-power by establishing visible and socially significant markers as well as by employing special mechanisms such as language, reasoning, and organization (Riles, 2011).

Conceptualizing persuasive power requires recognizing when power moves beyond the terrain of the sovereign while at the same time relying on the legitimacy of that sovereign. Persuasive power can operate by leveraging values and reason to will both behavioral and cognitive acceptance, without an explicit stamp of approval by the sovereign and, therefore, the requisite legitimacy via formal or legal structures of power. Persuasive power is influential in contemporary natural resource development on private lands in the US, particularly in rural, marginalized communities, with significant impacts for social justice and human health. Such development typically mixes assemblages of private and public, sovereign and corporate. The case of securing a lease to develop unconventional oil and gas resources provides one empirical example of persuasive power in action.

Acquiring rights for oil and gas resources

In the US, mineral resources can be owned by a private person or entity or municipal, state, or federal government (Sidortsov and Sovacool, 2015). Regardless of the resource ownership, the rights to explore, develop, and extract oil and gas are acquired pursuant to an oil and gas lease (Bosselman et al., 2010), which resembles a contract between two parties: the landowner, who can be a private or public actor, as lessor, and an oil and gas company or the "operator" as lessee (Bosselman et al., 2010). Whether the landowner is a public or private actor, an oil and gas lease is a contract into which the parties are supposed to enter freely. Thus, there should be no coercion or outright manipulation (Scott, 2001), with each party driven by ostensibly mutually beneficial economic arrangements in which the landowner leases their land and/or mineral rights in exchange for a signing bonus and potential royalty (Bosselman et al., 2010).

Private ownership of mineral resources contributes to the pace of unconventional gas development in the US (Sidortsov and Sovacool, 2015). Private ownership of a mineral resource and private agreement regarding its development does not mean that a sovereign is completely absent from the relationship created by an oil and gas lease. From the enforcement of a lease through courts to pooling requirements imposed by state conservation commissions and land use and zoning regulations adopted by some municipalities, sovereign power remains integrally involved in oil and gas development on private land (Bosselman et al., 2010).

The chief purpose of the types of corrective influence channeled through law is protection of public interest. Yet natural resource development on private land is inherently shaped by a power imbalance, originating from the structural positions of both parties (or uneven bargaining power) and the ability to create a constructed sense of mutual agreement through the coercive force of persuasive power (Aghion and Hermalin, 1990). As the Guidelines clearly indicate, the choice by many landowners is rarely free. The persuasive power of the industry as articulated by the Guidelines clandestinely turns an oil and gas lease into a coercive arrangement. Furthermore, persuasive power operates by legitimizing private interests through an implied connection to sovereign expressions sanction, utilizing public faith in the sovereign's legitimate hold over issues of public good by suggesting that private interests are tied to the interests and legitimacy of the sovereign. Analysis of the Guidelines reveals several mechanisms through which persuasive power is channeled, including setting boundary limits on the conversational field, invoking and manipulating appeals to reason and appeals to values, capitalizing on technical ignorance and economic vulnerability, and linking to legitimated forms of sovereign power. Each operates as a tool for utilizing persuasive power to shape oil and gas lease acquisition processes.

Persuasive power as setting the limits of the conversational field

Steven Lukes (2005) claims that power is most effective when it is least observable. By setting the limits of the conversational field, landmen employ the process of legitimation to misdirect the landowners. Landmen are instructed to set the tone of their conversation with landowners by deciding what is acceptable to talk about, focusing on individual fears and points of agreement. One of the very first points made in the Guidelines acknowledges the "increasing resistance from local community groups, so it is essential to contact land holders and acquire signatures before sentiment by environmental and other public organizations limits our ability to obtain access to private land." In the face of potential resistance, the intent is to be first to set the frame, establish the parameters of the possible, and exclude certain concerns: "Do not feel you have to discuss every point and question… Successful landmen understand what points to focus on so the benefits outweigh the cons."

Landmen are also instructed on how to explain the lease term, describing this concern as one landmen "can alleviate with the right wording." The primary term is for five years, and the secondary term is usually for as long as gas is produced in paying quantities (Bosselman et al., 2010). Landmen are instructed not to disclose the secondary term and to simply state that the lease is a five-year commitment. If a landowner were to know that the lease could last as long as 40 years, they may be more hesitant and want more time to weigh their options. Setting the limits of the conversational field also involves playing on individual and community level fears.

The notes instruct landmen to start with statements that appeal to the patriotic tendencies of rural, Midwestern Americans, telling them that China is now the leading purchaser of oil, as "Fear of foreign encroachment is the biggest asset we have in selling our development strategy."

Described as a "no-win discussion point," landmen are instructed to avoid the topic of property values because "studies have shown that property values decrease for land with oil and gas leases on the property" and "some banks have stopped issuing mortgages on properties with leases for mineral and oil/gas rights." If asked, landmen are told to redirect the conversation to focus on helping to meet the growing energy demand, their potential royalties, and back to the point about China. General fears, landmen are instructed, should be also reiterated any time the conversation turns to local or environmental impacts, so as to "re-direct the conversation… to the mutually agreed upon point about freeing the nation." As the notes instruct, "pure psychology… ALWAYS start your conversation with a new potential signee on a point that they will agree with." Beginning with points of agreement makes it more difficult to tease out the nuances of disagreement later; it also allows conversations to continue, which landmen are told increases the chance of signing.

Even what type of natural resource the company plans to use the lease for is not fully disclosed. Landmen are instructed to tell landowners the lease will primarily be used for oil exploration and to minimize any link to gas exploration, greatly stressing this to avoid any correlation to the movie *Gasland*. Additionally, the notes warn landmen of recent pushback due to natural gas's exemptions from various pieces of environmental laws, stating: "while this point is true for the exploration of natural gas, once again stress that we are searching for oil. Draw those lines clearly." Misinformation is used as a source of power to minimize unfamiliarity.

The Guidelines make it clear that landmen are even instructed to set conversation boundaries through directly misleading statements or clear deception. The notes tell landmen to avoid discussing the difference between hydraulic fracturing and slick water hydraulic fracturing; for those landowners aware of the difference, landmen should not mention water contamination or chemicals and are instructed to "reassure landowners that no well contamination has ever been documented. Do not mention water contamination in Pennsylvania." The notes explain, "after we have the leases signed we have the freedom to use the land as we see fit."

Persuasion occurs through selectively providing information, using rhetorical and discursive tools to shape the tone and content of the conversation. Further, persuasion relies on but is not synonymous with the structural relations of power that shape social dynamics. Landmen can rely on their structural position as perceived experts and on their tie to structurally powerful economic groups, but this structural position alone is not the source of their power; their power lies in persuasion.

Persuasive power through appeal to reason and values

Persuasive power can also be utilized via appeals to reason, including reason via scientific and technical facts that may require expertise or may be presented in a way that takes advantage of perceived technical ignorance. For example, the Guidelines state: "residential owners will not know that we pull water directly from the local aquifer" and that "radioactivity comes from natural sources in the ground and is released by the process, but don't tell them this." The company redefines this issue of radioactivity release as a natural process not related to fracking, making it seem that any potential concern regarding radioactivity is due to a homeowner's lack of knowledge on the subject (Davis and Hoffer, 2012). This manipulation of the truth and careful use of wording hinders understanding and works to wrongfully suggest that operations are safe.

An unconventional oil and gas drill site can occupy up to 20 acres of land; once complete, the well pad encompasses approximately five acres of land. The Guidelines instruct the landmen to focus on the five acres. Furthermore, in discussing well spacing, landmen are instructed to keep landowners from knowing that multiple wells may be required. If a landowner does inquire, they are instructed not to answer that question, explaining, "most landowners will not realize that 10–20 wells can be placed in a square mile." This creates an inaccurate message to the landowners in an attempt to minimize their concerns regarding potential impacts to their land (Hudgins and Poole, 2014).

An appeal to reason based on persuasive power also involves invoking economic rationality and vulnerability. Landmen are instructed to address concerns about increased truck traffic by convincing landowners to think about it as "the more trucks, the more royalties," persuading landowners that the extra money will outweigh the truck traffic. Similarly, the notes suggest landmen bring up potential revenues if forced into the topic of property values, a discourse used to persuade the landowners, not edify them (Hudgins and Poole, 2014).

Persuasive power can also be employed via an appeal to social values and agreement among rural Midwestern Americans regarding the value of small business. Landmen are instructed to talk about the land acquisition process as part of a small business, although noting that the lease will likely be sold to a larger company. By focusing on this enterprise as a small business, potential signees may associate particular values with the business; the power of persuasion operates by suggesting potential associations between structures and corresponding social values (even when they do not exist) and allowing those mental associations to form even without direct expression of their basis in reality.

Private landowners, particularly those located in rural America where natural resource development is dominantly based, may also share particular values associated with home, including the ideal imagery of a quiet refuge. Oil and gas development can disrupt rural communities through noise, traffic, and other infrastructural and social changes (Adgate et al., 2014). For example, compressor stations, located on each well pad, emit 85–95 decibels 24 hours a day, even though Occupational Safety and Health Administration (OSHA) regulations only allow noise levels in this range for an eight-hour period (Christopherson and Rightor, 2012). The Guidelines explicitly suggest making landowners think noise will only be during the construction of the site. If the landowner inquires further, the notes tell landmen to say the company closely monitors the noise levels to ensure they do not exceed 80 decibels at 200 feet, with instructions to "stick with the numbers, they provide the truth but make it hard to understand the exact implication."

Landmen are instructed to use aspects of a community's identity to leverage persuasion. They are told that the typical resident they address "will welcome you into their home and allow you to speak" and that "mid-western Americans dislike confrontation." The Guidelines urge capitalizing on this cultural characteristic, emphasizing that it is "critical to obtain a lease signature in the first meeting" and even to drive the person to the notary if necessary.

The final point commands landmen to "Get the lease signed!" For landowners who are undecided at the end of the visit, landmen can offer additional incentives to try to close the deal. The first option involves extra compensation for the landowner's first lease payment; the notes say even offering $50 to cover the notary costs can be effective in getting the landowner to sign. The second option is for landmen to say "all their neighbors have signed" even if they have not. Moreover, when the landowner does sign, the notes praise this as an opportunity for the landmen to "show those leases to undecided neighbors for added pressure." The third option, and final recourse, is to offer the landowner an amendment to the lease that will include a phrase that prohibits the use of slick water hydraulic fracturing techniques. The most important part is having a lease signed so that the company can complete the initial drilling and assess the potential productivity of the well. Presenting these financial incentives, making false claims about neighbors, and offering amendments that can be easily modified, put landowners under an overwhelming pressure to sign. In addition, using technical language creates an illusion of the landmen as experts in the field with professional legitimacy (Hillier, 1993).

Persuasive power through implied sovereign legitimation

Persuasive power works through implied associations with professional legitimacy, as well as with the legitimacy of the sovereign. To further reassure landowners, the Guidelines mention that the State Department of Environmental Quality (DEQ) continues to approve new drilling permits and "would not do so if the process were unsafe." By mentioning the DEQ this way, landmen suggest legitimated power, suggesting to landowners that they have the support of the state. This raises concerns regarding states' relationships with the industry, as natural gas companies and state political authorities have both benefited from state control of the industry.

This invoking of connections to sovereign legitimacy is perhaps the most significant element of persuasive power, in that it implies ties to structural relationships of power through construction, discourse, and even manipulation. Landmen are instructed to begin conversations about potential lease contracts with statements that appeal to the patriotic tendencies of rural, Midwestern Americans, statements about the need for a stable domestic source of energy, statements about China. Additionally, the conversation should be redirected to these points if landowners raise concerns about property values. Landmen tie private corporate interests to sovereign interests of the public good. By making this connection and drawing on this purported fear of foreign encroachment, landmen are utilizing persuasive power by employing a constructed relationship between the private gain they are pursuing and the public interests that might appeal to individual landowners. They are utilizing implied support from structural relations of power to construct a discursive field in which their interests become implied public interests. Persuasive power operates to discursively connect the private interests in development to public interests managed for the public good, such as the clear connect to sovereign power implied in raising "the mutually agreed upon point about freeing the nation."

Persuasive power and the challenge for communities

Legitimation can be upheld via dictating the boundary limits of the conversation and by keeping selected information or potential problems a secret. Legitimation may be sustained by misdirecting the public and keeping them focused on other topics (Freudenburg and Alario, 2007). In the case of natural resource development, persuasive power is a particularly problematic dynamic. High-volume hydraulic fracturing is negatively impacting rural communities across America (Ridlington et al., 2016). Attentiveness to the geographies of energy development illustrates how natural resource development for energy production can reshape rural landscapes (Bridge et al., 2013). Areas with extractive industries are sometimes referred to as a "zone of national sacrifice" where the health and well being of the land and community are sacrificed as a result of resource commodification (Hudgins and Poole, 2014, p. 307). The federal and many state governments do little to directly protect private well water, leaving it up to the well owner themselves to ensure its safety (Perry, 2012).

According to a Congressional Report on the substances used in fracking fluids, over "2,500 hydraulic fracturing products containing 750 chemicals and other components" were used between 2005 and 2009 (US House of Representatives, 2011, p. 1). Furthermore, "one or more of 29 chemicals that are (1) known or possible human carcinogens, (2) regulated under the Safe Drinking Water Act for their risks to human health, or (3) listed as hazardous air pollutants under the Clean Air Act" were found in over 650 different fracking products (US House of Representatives, 2011, p. 1). In addition to physical health, natural resource development is associated with the boom and bust cycle, which can cause social impacts including increased stress, diminished community cohesion, and changes to community character (Boudet et al., 2014).

While natural resource development is more likely to involve and impact rural spaces, individuals living in rural communities may be especially vulnerable to persuasive power because they are more likely to be trustful of their neighbors (Gao, 2016). Rural communities may perceive themselves as similar to a rural idyll (Halfacree, 1994) where people live an unhurried, healthy, and happy life in a close-knit community with a nearby abundance of natural resources, and thus may not perceive the calculated risks involved in natural resource extraction, as they imagine their communities as safe, protected places. Rural places are often cognitively associated with perceptions of naturalness (Hands and Brown, 2002), which may also limit the ability of rural residents to imagine a future for their communities in which natural resource extraction activities have created industrial or post-industrial landscapes and harmed the health of residents. Persuasive power exploits both the materiality and the shared social meanings associated with place to shape natural resource development in ways that benefit private interests, with a presumed but not necessarily clear linkage to public goods.

Addressing persuasive power

Attentiveness to how persuasive power operates is essential for uncovering these processes of legitimation. By understanding how processes of legitimation work, landowners and communities will be better prepared if approached about natural resource development opportunities. Persuasive power distorts possibilities for considered approaches to the public good of natural resource use through implied sovereign legitimacy, inequitable balance in limiting discourse, and appealing to socially constructed reason and values; recognizing these mechanisms implies a means of addressing them through concerted appeal to balanced access to knowledge of public concern for sustainable resource management and healthy communities.

Persuasive power manipulates cognitive perceptions through setting the boundaries of the conversational field and appealing to both reason and value through presumed technical ignorance, economic vulnerability, and perceived social positions. It is arguably most effective because it discursively links private interests with sovereign legitimacy, creating constructed and manipulated connections between pursuit of private gain and public interests. Private business entities will use persuasive power to pursue their interests, even in legally ambiguous ways, because their responsibility is to increase owner profits (Hansmann and Kraakman, 2000). However, these entities utilize presumptions about the private and mutually agreed upon nature of private contracts by appealing to presumed sovereign legitimacy and connection to issues of public good. Thus, the means of addressing persuasive power must come through the direct support of the sovereign to tease out, address, and ultimately combat persuasive power for the sake of protecting public health and well being.

Hydraulic fracturing's exemptions from provisions in several environmental laws at the federal level shifts regulatory responsibility to states (North et al., 2014). This raises concerns about potential impacts as well as questions about how local governments can best protect their communities (Rawlins, 2014). The US Supreme Court in Penn Central Transportation Co. v. New York City (1978) acknowledged an assurance under the Fifth Amendment that is "designed to bar Government from forcing some people alone to bear public burdens which, in all fairness and justice, should be borne by the public as a whole." If applied to the natural gas industry, the burden might be placed on the industry to provide greater transparency and evidence of safety, rather than on the personal landowners or community (Rawlins, 2014).

Rachel Kitze argues "that the current fracking regulatory system is failing to protect the interests of local communities and that local governments must retain meaningful control over fracking because they are in the best position to understand how it affects their communities" (Kitze, 2013, p. 387). Local governments should work to combat persuasive power by adding regulations at the local level and improving the oversight and enforcement of the industry operating in their communities (Rawlins, 2014). One example of a seemingly successful program is the Center for Responsible Shale Development (CRSD), a non-profit organization comprised of a variety of stakeholders, including environmental and philanthropic groups, industry representatives, and prior politicians. The CRSD has developed performance standards as well as a certification process and monitoring program overseen by an independent third party, with the goal to protect communities in the Appalachia Basin (CRSD, 2016).

One challenge is the very nature of private contracts for environmental resource acquisition; when companies obtain oil and gas leases from private owners, it is an agreement made only between the landowner and the company (North et al., 2014). However, these private agreements have incredible consequences for the health of the communities and citizens where they would operate. Local governments and community groups can empower private landowners by providing them with a basic guide on a typical oil and gas lease and by connecting them to resources on what they should do if approached by a landman. This would help prepare and educate landowners so that they might be less prone to the persuasive power used by landmen.

In instances where persuasive power interferes with sustainable resource management in ways that negatively impact public health or the public good, sovereign power can take an active role in rearticulating the relationship between private and public interests so as to redress the power of persuasion that favors the advancement of private interests above public goods. Specific modification to processes of private party contracts could involve, for instance, instituting mandatory 24- to 48-hour "mull-over" periods (that can be waived by especially eager landowners) to help tamper persuasion through emotional responses to value appeals.

State governments could consider acting in an advisory role and serving as a resource. Such a designated unit can be hosted in a state agency responsible for consumer protection. The unit can offer general consultations on the applicable legal and regulatory framework, clarify risks associated with oil and gas development, and provide additional data to landowners. Providing landowners with resources, including knowledge resources, could help restore information balance. Community-scale decision-making can be supported through a multiplicity of mechanisms, all of which could operate to address persuasive power by clearly articulating the role of the sovereign in legitimating and managing natural resources for the public good and supporting the ability of all perspectives to contribute to deliberation regarding reasons and values. This could be done directly by agents of the sovereign, consumer protection agencies, community advocacy organizations, or other non-governmental advocacy groups.

Conclusion

Persuasive power occurs when one pushes beyond the boundaries of sovereign power while also depending on the legitimacy of that power. The use of persuasive power plays an influential role in shaping natural resource use, with profound impacts for human health. Landmen provide a good example, with their strategic control of the conversation, careful concealment of information, points of persuasion based on commonly shared reason and presumed social values, and assumed legitimation tied to professional, technical, and a presumed link to sovereign power. Therefore, persuasive power negatively impacts human health by influencing landowners; this one example demonstrates how persuasive power poses a challenge to managing natural resources to protect the health of communities.

Given that the most significant tactic utilized in the deployment of persuasive power is the implied linkage between private interests and sovereign legitimacy, combating persuasive power in order to develop models for sustainable and community-driven natural resource development ought to come from the terrain of sovereign power. By creating consumer advocacy networks and channels for public deliberation, sovereign power can be used to differentiate among public goods and the risks and benefits facing different publics, on the one hand, and the interests of private entities on the other. These private interests in natural resource development are not in themselves granted sovereign legitimacy, but persuasive power creates discursive links that disguise their interests and the scope of existing granted power to operate in legally legitimized ways. Thus, the sovereign must disrupt these links, engage publics, and establish practices that work to diminish the power of persuasion in colonizing public interests via the challenges of sustainable natural resource development and fostering healthy communities.

References

Adgate, J. L., Goldstein, B. D., and McKenzie, L. M. (2014). Potential public health hazards, exposures and health effects from unconventional natural gas development. *Environmental Science & Technology*, 48(15), pp. 8307–8320.

Aghion, P. and Hermalin, B. (1990). Legal restrictions on private contracts can enhance efficiency. *Journal of Law, Economics, & Organization*, 6(2), pp. 381–409.

Archive (n.d.). Talking Points for Selling Oil and Gas Lease Rights [online]. Archive. Available at: archive.org/details/TalkingPointsForSellingOilAndGasLeaseRights [Accessed 13 November 2016].

Boudet, H., Clarke, C., Bugden, D., Maibach, E., Roser-Renouf, C., and Leiserowitz, A. (2014). "Fracking" controversy and communication: Using national survey data to understand public perceptions of hydraulic fracturing. *Energy Policy*, 65, pp. 57–67.

Bosselman, F., Eisen, J. B., Rossi, J., Spence, D. B., and Weaver, J. L. (2010). *Energy, Economics and the Environment: Cases and Materials*, 3rd edition. Foundation Press.

Bourdieu, P. (1987). The force of law: Toward a sociology of the juridical field. *Hastings Law Journal*, 38, pp. 805–853.

Bridge, G., Bouzarovski, S., Bradshaw, M., and Eyre, N. (2013). Geographies of energy transition: Space, place and the low-carbon economy. *Energy Policy*, 53, pp. 331–340.

Castells, M. (2011). *Communication, Power and the State in the Network Society* [online]. University of Cambridge. Available at: www.youtube.com/watch?v=9-ubCGAV3iw [Accessed 1 August 2017].

Center for Responsible Shale Development (CRSD) (2016). CRSD's Official Website [online]. Available at: www.responsibleshaledevelopment.org [Accessed 15 November 2016].

Christopherson, S. and Rightor, N. (2012). How shale gas extraction affects drilling localities: Lessons for regional and city policy makers. *Journal of Town and City Management*, 2(4), pp. 1–20.

Davis, C. and Hoffer, K. (2012). Federalizing energy? Agenda change and the politics of fracking. *Policy Sciences*, 45(3), pp. 221–241.

Freudenburg, W. R. and Alario, M. (2007). Weapons of mass distraction: Magicianship, misdirection, and the dark side of legitimation. *Sociological Forum*, 22(2), pp. 146–173.

Gamson, W. S., Croteau, D., Hoynes, W., and Sasson, T. (1992). Media images and the social construction of reality. *Annual Review of Sociology*, 18, pp. 373–393.

Gao, G. (2016). Americans divided on how much they trust their neighbors [online]. Pew Research Center. Available at: www.pewresearch.org/fact-tank/2016/04/13/americans-divided-on-how-much-they-trust-their-neighbors/ [Accessed 13 November 2016].

Grewal, D. S. (2008). *Network Power: The Social Dynamics of Globalization*. New Haven and London: Yale University Press.

Halfacree, K. H. (1994). The importance of "the rural" in the constitution of counterurbanization: Evidence from England in the 1980s. *Sociologia ruralis*, 34(2–3), pp. 164–189.

Hands, D. E. and Brown, R. D. (2002). Enhancing visual preference of ecological rehabilitation sites. *Landscape and Urban Planning*, 58(1), pp. 57–70.

Hansmann, H. and Kraakman, R. (2000). The end of history for corporate law. *Georgetown Law Journal*, 89, p. 439.

Hillier, J. (1993). To boldly go where no planners have ever... *Environment and Planning D: Society and Space*, 11(1), pp. 89–113.

Hudgins, A. and Poole, A. (2014). Framing fracking: private property, common resources, and regimes of governance. *Ecology*, 21, pp. 222–348.

Kitze, R. A. (2013). Moving past preemption: enhancing the power of local governments over hydraulic fracturing. *Minnesota Law Review*, 98, p. 385.

Kreuze, A., Schelly, C., and Norman, E. S. (2016). To frack or not to frack: Perceptions of the risks and opportunities of high-volume hydraulic fracturing in the United States. *Energy Research and Social Science*, 45, pp. 45–54.

Ladd, J. M. and Lenz, G. S. (2009). Exploiting a rare communication shift to document the persuasive power of the news media. *American Journal of Political Science*, 53(2), pp. 394–410.

Lukes, S. (2005). *Power: A Radical View*. 2nd edition. Basingstoke and New York: Palgrave MacMillan.

Morris, P. (2002). *Power: A Philosophical Analysis*. 2nd edition. Manchester: Manchester University Press.

North, D. W., Stern, P. C., Webler, T., and Field, P. (2014). Public and stakeholder participation for managing and reducing the risks of shale gas development. *Environmental Science & Technology*, 48(15), pp. 8388–8396.

Penn Central Transportation Co. v. New York City, 438 U.S. 104 (1978).

Perry, S. L. (2012). Addressing the societal costs of unconventional oil and gas exploration and production: A framework for evaluating short-term, future, and cumulative risks and uncertainties of hydrofracking. *Environmental Practice*, 14(4), pp. 352–365.

Rawlins, R. (2014). Planning for fracking on the Barnett Shale: soil and water contamination concerns, and the role of local government. *Environmental Law*, 44(1), p. 135.

Ridlington, E., Norman, K., and Richardson, R. (2016). Fracking by the numbers: The damage to our water, land, and climate from a decade of dirty drilling [online]. Environment America Research & Policy Center: Frontier Group. Available at: www.environmentamerica.org/sites/environment/files/reports/Fracking%20by%20the%20Numbers%20vUS.pdf [Accessed 15 November 2016].

Riles, A. (2011). *Collateral Knowledge: Legal Reasoning in the Global Financial Markets*. Chicago and London: University of Chicago Press.

Russell, B. (1938). *Power*. London: George Allen & Unwin Ltd.

Scott, J. (2001). *Power*. Cambridge and Malden: Polity Press.

Sidortsov, R. and Sovacool, B. (2015). State–market interrelations in the US onshore and offshore oil and gas sectors. In: Belyi, A.V. and Talus, K. eds., *States and Markets in Hydrocarbon Sectors*, Palgrave Macmillan UK, pp. 171–197.

US Energy Information Administration (EIA) (2016). Shale in the United States [online]. Available at: www.eia.gov/energy_in_brief/article/shale_in_the_united_states.cfm [Accessed 13 November 2016].

US House of Representatives, Committee on Energy and Commerce (2011). Chemicals used in Hydraulic Fracturing [online]. Available at: democrats.energycommerce.house.gov/index.php?q=news/committee-democrats-release-new-report-detailing-hydraulic-fracturing-products.

Weber, M. (1978). *Economy and Society*. Berkley, Los Angeles, and London: University of California Press.

9

SOCIAL AND POLITICAL INEQUALITY AS CHALLENGES IN TECHNOLOGY DIFFUSION

Evidence from government-funded improved cookstove program in rural Mexico

Aparajita Banerjee and Chelsea Schelly

Envisioning a sustainable world requires not only a transition to sustainable socio-technical systems (eg, energy, mobility, agriculture), but also a "fundamental change in the structure, culture, practices" (Rotmans and Loorbach, 2010, pp. 108–109) in societies. Such alterations of societal regimes are not easy, and often there is resistance to change as it contests the status quo held in place by dominant actors, structures, and practices of the current regime. Though policies can play a critical role, there is also need to understand societal power dynamics and plan accordingly by understanding how existing power relations can resist or create hindrances in any plans to transition. As one theme of this book is to have a better understanding of the challenges facing environmental policies in the quest for sustainability, this chapter presents a case to show how policy programs for technological transition face inherent problems related to social and political inequality, which affect technology diffusion and ultimately the pursuit of sustainability.

In the next section, we describe in detail the concept of power in the context of social and political inequality and discuss social dominance theory to provide a theoretical explanation of our study findings. We then provide a brief overview of the critical role of clean cookstove technologies for achieving global sustainability and sustenance. Then we present our case study and briefly describe the research methods followed by the results. A discussion followed by a conclusion section recapitulates our findings in the context of the central themes of this book.

Unpacking inequality and social dominance theory

Power is one of the most contested concepts in social and political theory, and there are a number of ways power has been understood and defined. Power is the social relations between two types of agents: the principal and the subalterns, where the principal holds or exercises power as the dominant agent, which in turn affects the subaltern or the subordinate agent (Scott, 2001). The principal(s) can transform the subaltern in such a way that the latter behaves conforming to the expectation of the former even without the principal(s) exercising force or imposing constraints (Lukes, 1974). This can be done by way of how the principal(s) distribute benefits and burdens to the subalterns (Dowding, 1996). In doing so the principal(s) can use manipulation like providing benefits, resources, and access to opportunities to the subalterns differently (Scott, 2001). Established social structures like the state can stabilize the power of the principal by providing a means of using different state institutions to exercise their power regularly, and members who belong to regional or local divisions of state also form local elite groups (Scott, 2001). Therefore, the principal(s) can distribute resources to the subaltern to manipulate the subaltern or create discrimination between groups of subalterns by disproportionally sharing benefits and burdens. Not surprisingly, studies have found that local elites can control how the benefits from certain national or international projects reach non-elite subalterns (Acemoglu et al., 2014; Dasgupta and Beard, 2007). Others have found that benefits of government-initiated projects can also benefit the village elites disproportionately due to the presence of elite control (Lucas, 2016; Standing and Hilson, 2013), exacerbating conditions of social inequality and intergroup differences.

But why do the principal(s) use such manipulative tactics? The principal(s) do so because even though the subaltern may or may not exercise their power, they have the potential to do so and to affect the principal(s) due to the relational constitution of the social context in which they both exist (Barnes, 1988; Clegg, 1989). Therefore, the principal(s) manipulate loyalty and use discriminatory tactics by favoring one group over another subaltern group to decrease potential threat from them. Social dominance theory (SDT) helps elucidate this in further detail by explaining intergroup discriminatory behavior between two social groups (Sidanius and Pratto, 2011). SDT posits that the existence of such intergroup hierarchies can lead to intergroup oppression where "members of the dominant group secure a disproportionate share of the good things in life" and the non-elites or subalterns receive a "disproportionate share of the bad things in life" (Ibid., p. 418). SDT suggests that intergroup discriminatory behavior is often justified using a hierarchy-enhancing myth that legitimizes intergroup oppression. Social institutions often perpetuate such social hierarchies by allocating resources in ways that benefit the dominant group and become a burden for others; SDT terms these institutions "hierarchy-enhancing social institutions" (Ibid., p. 419).

Two processes contribute to perpetuating intergroup inequalities resulting in disproportionately allocating benefits and burdens. Intergroup discrimination between social groups involves temporal consistency and continuity based on, among other things, "remembered histories of past conflicts" and "perceived intergroup threats" (Ibid., p. 419). Individuals can also perpetuate social hierarchy-enhancing attitudes and behaviors based on their desire to be part of the group that can dominate others, which SDT terms as the social dominance orientation (SDO) (Ibid.). Therefore, the principal(s) indulge in discriminatory behavior in order to retain the loyalty of their favored group of subalterns or to woo other groups of subalterns to become part of the principal's favored group.

The above explanation leads us to the key question at the heart of this chapter: how does social and political inequality affect pursuit of sustainability? We use a case study approach to show how political and social inequality can play a critical role in the success of programs intended to improve environmental conditions. Using results from qualitative research in two communities in Yucatan, Mexico, where government-initiated Improved Cookstove (ICS) dissemination programs were implemented in 2016, we argue that social and political inequality in rural communities leads to elite control of government-sponsored distribution projects and plays a critical role in the implementation and successful uptake of ICS. Findings suggest that the ICS programs benefited certain groups of people affiliated with the dominant political party or close to the local leaders of the party; in many cases, these adopters did not have any real use for the ICS. We use SDT (Sidanius and Pratto, 2001) to explain these findings, arguing that tackling corruption and bypassing the ills of principal vs. subaltern power relations is critical to ICS dissemination programs and other attempts to address technology adoption via incentive-based policies and programs. These findings not only contribute to identifying local-level social and political challenges in implementing sustainable projects but also draw attention to critical social hierarchy issues that need to be addressed while developing implementation plans for future sustainable development programs or climate change mitigation and adaptation programs.

Improved cookstove programs worldwide

According to the World Energy Outlook 2016, 2.7 billion people, or around 38% of the total world population, continue to use biomass as cooking-fuel in traditional and inefficient cookstoves (IEA, 2016). As a result, indoor air pollution (IAP) from biomass combustion continues to be one of the foremost contributors to global disease burden (Lim et al., 2012). In addition to localized human health impacts, IAP significantly contributes to human-induced carbon emissions that impact the climate system (Bond et al., 2013). As such, many international and national agencies have introduced programs to facilitate the use of ICS in rural areas worldwide, technologies specially designed for increased fuel efficiency and lower emissions (Barnes et al., 2012; Venkataraman et al., 2010).

Since the 1970s when international development agencies started distributing ICS, mostly as a response to the energy crisis, a significant amount of research has been done to identify the benefits (Ruiz-Mercado et al., 2011; Bielecki and Wingenbach, 2014). ICS have health benefits (Boy et al., 2000; Kyu et al., 2010), can improve local ecosystems by reducing the need for biomass extraction, and can reduce global greenhouse gas (GHG) emissions (Lewis and Pattanayak, 2012; Nepal et al., 2011). Existing research has also linked ICS adoption with benefits related to gender dynamics and gender inequalities in the household (Gill et al., 2010). Fuel efficiency characteristics of ICS can help reduce household expenditure on domestic fuel (Modi et al., 2005), thus freeing up limited domestic income for other activities. Rural women and children are often in charge of collecting and gathering firewood, which restricts participation in other economic activity, exacerbating time-poverty issues (Gill et al., 2010). Increasing fuel efficiency can mean less time spent in fuel-gathering activities (Parikh, 1995) thus freeing up time for women and children to engage in learning new skills and pursuing formal education, ultimately leading to increased economic opportunities and overall empowerment (Modi et al., 2005). The health benefits of ICS can also improve the economic condition of rural women by reducing unproductive sick days (Venkataraman et al., 2010). Therefore, ICS adoption in rural areas of the peripheral world can contribute to sustainable rural conditions.

However, widespread ICS adoption is riddled with challenges arising from variants of social, political, or economic inequality. Economic inequality in rural areas makes ICS adoption less lucrative than in urban areas, as rural community members attach less opportunity cost to their time spent in fuel collection (Damte and Koch, 2011). Others report that ICS adoption has been successful when multiple methods are used, like creating conditions for credit access with different rates for different groups of people, subsidy programs, and methods of repayment (Barnes et al., 2012; Osei, 2010). Research also suggests that ICS programs where women participated in the development of ICS were more successful (Jin et al., 2006; USAID/Winrock, 2009). Others have found that women belonging to the traditionally patriarchal system and having lesser power in domestic decision-making either saved money from other expenses to pay for ICS or could negotiate such purchase with the help of other female family members (Muneer, 2003; Person et al., 2012).

Notably, most of the above literature dealing with equity and equality considerations in ICS dissemination focuses on how such social conditions impact ICS diffusion when people pay for the cookstoves. Yet non-market implementers, including non-governmental organizations and local-level government officials, implement many ICS programs worldwide (Smith et al., 1993; Troncoso et al., 2007). Though recent work on economic development and natural resource management has identified the phenomena of elite control of program implementation in rural communities (Acemoglu et al., 2014; Armitage, 2005; Dasgupta and Beard, 2007), this problem has not been discussed in ICS implementation literature. We aim to fill this gap as well as use our results to suggest the challenges social and political inequality creates for effective environmental policy implementation.

Materials and methods

Around six million people in Mexico continue to use firewood as a primary or secondary source of fuel (GACC, 2013). In rural areas, firewood is used mostly in open fires with stones surrounding the fire on an indoor kitchen floor or an outdoor setting that works as a kitchen. There have been efforts in Mexico to promote ICS technologies since the 1970s (Troncoso et al., 2007). The success of these programs has been mixed, and others suggest this is because the ICS were developed based on technical considerations rather than adapting to users' preferences (Ruiz-Mercado et al., 2011). In 2007, the Social Development Ministry (SEDESOL), a federal agency, introduced a national program for ICS dissemination targeting the 500 poorest municipalities in the country, where the use of traditional cookstoves is most common (López-Martínez and Cuanalo de la Cerda, 2017). As setting national GHG reduction targets gained importance in recent years, these programs have been identified as critical to reducing GHG emissions in Mexico (GACC, 2013).

In 2016, the SEDESOL launched the Program for Housing Improvement in the Modality of Ecological Stoves (English translation of *Programa de Mejoramiento de Vivienda en la Modalidad de Equipamiento de Estufas Ecológicas* or the Mejorar Program in short) and worked with the state government of Yucatan to distribute around 3,000 ICS (commonly known as *estufas ecológicas* or ecological stoves) as part of the government initiative (Diario Oficial, 2016). The Secretary of Social Development of the State Government of Yucatan was responsible for the implementation of the program. The official policy document identified that around 40.4% of Yucatan's population live in poor housing conditions and use traditional firewood stoves for cooking (Ibid.). Therefore, the objective of the program was to help create a better living space in marginalized rural Yucatan households, as guaranteed by the Mexican Constitution's Article 4 that recognizes that every family has the right to enjoy a decent living (Ibid.). Marginalized people defined by the program documents were disabled individuals and people who could not access nutritious food even when they completely dedicated their time and effort to its acquisition (Ibid.).

This study aimed to assess the Mejorar Program's effectiveness in providing ICS to rural people, especially to the most marginalized, fulfilling the primary objective of the program. We selected two villages in two municipal districts in Yucatan that vary in their social, economic, and political characteristics even though both are rural communities. The first is Sierra Papacal, situated 32 km north of the local urban center of Merida. Around 986 people lived here in 251 houses (INEGI, 2010). The houses are permanent structures differing in size and number of rooms, and most have access to public water and electricity services (Ibid.). Public transport connects Sierra Papacal with Merida. Thus, many people commute to Merida to work in low-skilled jobs like gardening, painting, construction, and domestic help.

The second village examined in this study was the Maya village of Sucopo in the municipality of Tizimin, with a population of 1,517 inhabitants (Sacramento Rivero et al., 2016). The community has a high degree of marginalization with elevated levels of poverty, low rates of formal education, and low average domestic income (INEGI, 2010). People live in either permanent houses or Mayan wood houses with roofs made from palm leaves. The community is 10 km away from the local municipal town of Tizimin, but there is no public transportation. The village has a community center, a health center, and a primary school. Community members are mainly subsistence agricultural producers or work in local low-skilled jobs. Members of many households in the villages also migrate to local tourist centers like Cancun and Playa del Carmen for jobs, occasionally visiting family. Many other members of the community travel for seasonal jobs like octopus fishing or plantation jobs.

This study utilized qualitative research methods including documentary research, participant observation, and semi-structured interviews for data collection. Documentary research involved reviewing policy articles and media reports on the cookstove program in local newspapers. A case-study research design approach included spending a total of 15 days in both communities and interviewing a total of 50 women in the two communities using a standard protocol. Interviewees were selected systematically. In Sierra Papacal, the 30 participating interviewees were selected from a total of 251 houses. Using a village map, the interviewer approached every eighth house to solicit an interview. The focus was on interviewing women and it was not difficult to find them at home most times. When the female member of the family who oversaw household cooking was not available during the first visit, the interviewer made an appointment or went back at a different time of the day. In Sucopo, 20 interviewees were selected from among 365 houses in the community using the same method as in Sierra Papacal, approaching every eighteenth house of the village to ask for an interview.

Results

All except one of our interviewees lived in self-owned houses, with an average household size of four people. Thus, they had a choice in building their kitchen to meet their needs, often impossible in a rented house. All interviewees had access to electricity in their houses. Interviewees had one of two types of kitchen arrangement: a place inside the house was designated as a kitchen where an LPG stove was used for cooking, or a traditional wood stove was outside the house, or they had both (see Figures 9.1 and 9.2).

We found that 27% of interviewees in Sierra Papacal and 50% of community members in Sucopo continued to use wood as the primary or only fuel for cooking. In these houses, there were no other types of fuel used for the kitchen. Other studies have suggested that 40.5% of rural people continued to use wood in Mexico until 2012–2013 (Hernández-Garduño et al., 2017). Interviewees either bought their wood from local vendors or asked a male member of their families to collect wood for them. Very few interviewees told us that they oversaw wood gathering and storage.

Not all community members were given an ICS by the government project. In Sierra Papacal, interviewees and the village headman informally told us that the ICS were given to 50 female members of the community selected across age groups, whereas in Sucopo, the ICS were given to roughly 200 male members of the community. Using our systematic selection method, we interviewed 12 women (40%) in Sierra Papacal who received an ICS from the government and ten women (50%) in Sucopo whose husbands got the ICS. Among the 12 women we interviewed in Sierra Papacal, 92% of the interviewees (11 out of 12) already had a wood stove and a gas stove when they were given an ICS. Similarly, in Sucopo, 70%

FIGURE 9.1 Domestic cooking arrangement using wood
Source: Aparajita Banerjee and Chelsea Schelly

of the houses that received the ICS already had both gas and wood for cooking. Only one interviewee in Sierra Papacal and three interviewees in Sucopo who used the traditional stoves exclusively received the ICS. According to the interviewees, someone from the municipal-level authority demonstrated the use of the stove in the village center before the distribution and each stove came with a pamphlet with instruction for setting up and using the ICS.

We found two distinct patterns of dissemination for the ICS or '*estufa ecologica*' in the two communities. To give context, it is critical to provide some background of the local politics. The village head in Sierra Papacal had been recently elected and is a member of the PAN (Partido Accion National). However, as Olvera (2016)

FIGURE 9.2 Domestic cooking arrangement using wood
Source: Aparajita Banerjee and Chelsea Schelly

has suggested for the whole of Mexico, the dominant political party in Mexico is PRI (Partido Revolucionario Institucional), which has a long history of ruling the country, retaining significant political and even cultural powers in the community, which was evident in the interviewee responses from Sierra Papacal. In Sucopo, the village headmen have traditionally been members of the PRI, and there has been no significant change in that status quo in recent years.

Among the 12 women in Sierra Papacal who received the ICS, 92% of the interviewees informed us that they got the ICS because they are a member of the PRI, and that the local party leader in their community informed them about getting a cookstove. They were asked to contact the village police station with their credentials, such as voter identity card, electricity bill, and birth certificate. One interviewee said:

> I heard people say that the government gave the stove... when I contacted the local party manager, they said that they requested the party for the cookstove and cookstoves were from the party.

Another interviewee said:

> Those ladies are from PRI who are block managers ... they told me to take my papers. I asked them if my mother can enter the program, but as my mother does not know her, they did not tell her anything about the program.

We also interviewed 18 interviewees in Sierra Papacal who did not get the ICS. Half of them (nine out of 18) informed us that they did not get the stove because the program was for members of the PRI, and they were not party and hence not eligible for the stoves. For example, one interviewee said:

> The program was for the PRI people... they submitted their PRI papers to get the stove ... the [people who are] PAN, or other different ones like that [political party], the program did not touch them.

Another interviewee said:

> What I got to know here was that, because they sent it for a specific party, only those people who belonged, they gave it to them ... the local party manager knows who is from the party and gave it to them.

Other interviewees from Sierra Papacal gave mixed responses on why they did not get the ICS. Some interviewees said that they did not know why they did not get the stove whereas others said that they had to choose one among different welfare programs and they opted to get vegetable seeds instead of the cookstoves.

In contrast, the village headman informed us that Sucopo has traditionally elected members of the PRI as the village headman and members of the same party have replaced the previous member. Community members here said that the ICS were given to the people whose names were on a list of community members.

Discussion

There continues to be a heavy dependence on using firewood for cooking in rural Yucatan and on woodstoves outside the house, similar to other studies in Mexico that suggest that there is a high use of firewood as cooking fuel (Berrueta et al., 2017; Masera et al., 2007). Using the cookstove outside the house can significantly reduce IAP impacts on health, as less particulate matter is collected than when the kitchen is inside the house. However, the heavy dependence on firewood as the only fuel for cooking continues the environmental pollution and ecological stress of firewood as cooking fuel. Therefore, ICS replacing woodstoves can improve ecological conditions as well as provide an efficient cooking option that saves money, time, and physical effort in fuel collection, all of which could be used in pursuing other social or economic activities that improve standard of living.

However, results also indicate that government-initiated projects differ in their implementation, which is often decided by local implementers. In the case of Sierra Papacal, ICS were given to women members of the community (who are responsible for daily cooking tasks and decision-making), whereas in Sucopo ICS were distributed to male community members. Other studies have shown that user participation in designing, training, and implementation in technology adoption is critical to program success (Gill et al., 2010). The difference in implementation can in turn affect the use of the cookstove and its sustainability impacts.

The findings most importantly point to the role of local-level politics in the implementation of ICS programs. We found in Sierra Papacal that though the local village headman belonged to a different political party from the national leadership, the dissemination of the cookstove program was mainly decided and controlled by the block chief of the party in power at the state level in Yucatan and at the national scale. The local-level members of the party at the center of Mexican politics (PRI) have become what Scott (2001) terms as local power elites. Their position has been strengthened in local politics, as PRI has been the authority in Mexico since the independence of the country to the present, apart from the 12 years between 2000 and 2012 (Olvera, 2016). Given such a long rule in the country by a single political party, the party has become an important institution and cultural influencer with substantial presence in rural communities (Olvera, 2016). Such a long history of power has aided in building what SDT terms a "hierarchy-enhancing" myth that can justify discriminatory behavior on the part of the dominant group on the subordinate group in the local society (Sidanius and Pratto, 2011). People who received the ICS told us they got the stove because they were members of PRI and others did not get it because they were not members of the party. Even people who did not belong to the PRI and perceived that they did not benefit from the program did not show any significant discontent in such discrimination, simply pointing out that the program was for the members of the PRI. In Sucopo, where PRI has been the dominant political party and there has been no other active representation of other political parties in village politics, ICS beneficiaries were predetermined without clear bias based on party affiliation.

The decision of the local party managers to not include non-party members in the ICS dissemination program in Sierra Papacal also can be explained by SDT, which suggests that intergroup discrimination can exist when the socially dominant group "perceive[s] intergroup threats" (Sidanius and Pratto, 2011, p. 419). Members of the dominant group can discriminate against the subordinate group if they see that any distribution of resources that can benefit one group and not the other will reduce risk from the threat they have perceived. Thus, in Sierra Papacal, providing

ICS to the members of the party was arguably used to sustain the critical loyalty of the party members to the party particularly in a time when such loyalties were threatened by the election of a different party in the local village election. The acts of the local block managers to seek people who support their party to benefit from the ICS dissemination program can be explained by what SDT terms the social dominance orientation, where the block manager acts to maintain group-based inequality.

Interviewees in Sierra Papacal also informed us that the local party managers approached people in their community whom they knew to be members of the party for ICS dissemination, pointing to the existence of local social institutions created by party members to provide benefits to its members. Such institutions allow party members to implement government-initiated welfare programs in ways that benefit local members of the party. SDT terms these institutions as hierarchy-enhancing institutions (Sidanius and Pratto, 2011, p. 419). Due to the presence of such discrimination in dissemination, ICS were distributed to people who already had other improved cooking technologies like LPG gas stoves, whereas those who did not have such improved facilities and depended solely on traditional cookstoves were left out of the program.

Local elite control of government welfare programs is not unexpected. Democratic processes are far from ideal in Mexico, with the legislature, judiciary, and electoral process being heavily controlled by the executive branch of the government (Fox and Hernandez, 1992). Exchanges of goods and services for political support that can mutually benefit the local party member as well as the community member who gets the benefit is fairly common in Mexico and often legitimized as political clientelism (Hicken, 2011; Hilgers, 2008; 2011). However, such exchange of goods and services is not targeted to meet the needs of the people who need or would even use the service.

Conclusion

The success of ICS dissemination depends on how effectively they can provide human health benefits, ecological benefits from reducing firewood use, and environmental benefits from decreased GHG emissions. The benefit is greatest when people who use firewood the most start using the ICS. However, this study found that local-level politics of intergroup discrimination in resource allocation reduced the impact of reaching the target population via implementation. Such discrimination in resource allocation can become institutionalized to such a degree that it becomes legitimate to people affected both positively and negatively by such an arrangement.

ICS implementation is an outcome of policies that envisage a transition to a sustainable world. Yet social and political inequality, especially related to who has the power and how it is exercised, has a variety of localized manifestations. These manifestations, in turn, can affect well-intentioned policies and can limit their positive impacts. Therefore, the onus lies not only in framing environmental policies that pursue sustainable outcomes, but also in creating implementation plans that

can maneuver elite control, especially in the rural areas in the peripheral world where attaining sustainability is challenging but of utmost importance to ensure social justice and a sustainable world. Elite control arising from social and political inequality that perpetuates social inequality in village societies is not just inherent to Mexican society, but is visible in many peripheral countries (Blair, 2000; Collins and Green, 1994; Rambaldi et al., 2006). Greater understanding through research regarding the role of local political elites and intergroup discrimination is needed to improve the implementation of environmental policy decisions and technology adoption programs aimed at environmental improvements. Finding ways to tackle such issues can also assist in achieving these goals in ways that address the most vulnerable population who lack social and political power so that they get the benefits of policies or programs to pursue sustainability.
Acknowledgements: this work was funded by the National Science Foundation's Partnerships in International Research and Education (PIRE) Program.

References

Acemoglu, D., Reed, T., and Robinson, J. A. (2014). Chiefs: Economic development and elite control of civil society in Sierra Leone. *Journal of Political Economy*, *122*(2), 319–368.

Armitage, D. (2005). Adaptive capacity and community-based natural resource management. *Environmental Management*, *35*(6), 703–715.

Barnes, B. ([1988]2002). The nature of power, in: Haugaard, M., ed. (2002) *Power: A Reader*, Manchester: Manchester University Press.

Barnes, D. F., Kumar, P., and Openshaw, K. (2012). *Cleaner Hearths, Better Homes: New Stoves for India and the Developing World*. Oxford University Press.

Berrueta, V. M., Serrano-Medrano, M., García-Bustamante, C., Astier, M., and Masera, O. R. (2017). Promoting sustainable local development of rural communities and mitigating climate change: the case of Mexico's Patsari improved cookstove project. *Climatic Change*, *140*(1), 63–77.

Bielecki, C. and Wingenbach, G. (2014). Rethinking improved cookstove diffusion programs: A case study of social perceptions and cooking choices in rural Guatemala. *Energy Policy*, *66*, 350–358.

Blair, H. (2000). Participation and accountability at the periphery: Democratic local governance in six countries. *World Development*, *28*(1), 21–39.

Bond, T. C., Doherty, S. J., Fahey, D. W., Forster, P. M., Berntsen, T., DeAngelo, B. J. … and Kinne, S. (2013). Bounding the role of black carbon in the climate system: A scientific assessment. *Journal of Geophysical Research: Atmospheres*, *118*(11), 5380–5552.

Boy, E., Bruce, N., Smith, K. R., and Hernandez, R. (2000). Fuel efficiency of an improved wood-burning stove in rural Guatemala: Implications for health, environment and development. *Energy for Sustainable Development*, *4*(2), 23–31.

Clegg, S. R. ([1989]2002). Frameworks of power, in: Haugaard, M., ed. (2002) *Power: A Reader*, Manchester: Manchester University Press.

Collins, C. and Green, A. (1994). Decentralization and primary health care: Some negative implications in developing countries. *International Journal of Health Services*, *24*(3), 459–475.

Damte, A. and Koch, S. (2011). Clean fuel saving technology adoption in urban Ethiopia. *Energy Economics*, *36*, 605-61.

Dasgupta, A. and Beard, V. A. (2007). Community driven development, collective action and elite capture in Indonesia. *Development and Change, 38*(2), 229–249.

Diario Oficial del Gobierno del Estado de Yucatan (2016). [online] Available at: www.yucatan.gob.mx/docs/diario_oficial/indices/2016.pdf [Accessed on 9 May 2017].

Dowding, K. (1996). *Concepts in Social Thought: Power*. Open University Press.

Fox, J. and Hernandez, L. (1992). Mexico's difficult democracy: Grassroots movements, NGOs, and local government. *Alternatives, 17*(2), 165–208.

Gill, K., Brooks, K., McDougall, J., Patel, P., and Kes, A. (2010). Bridging the gender divide. How technology can advance women economically. Washington, DC: International Center for Research on Women.

Global Alliance (GACC) (2013). Cookstove fact sheet. Global Alliance for Clean Cookstoves.

Hernández-Garduño, E., Gómez-García, E., and Campos-Gómez, S. (2017). Prevalence trends of wood use as the main cooking fuel in Mexico, 1990–2013. *Salud pública de méxico, 59*(1), 68–75.

Hicken, A. (2011). Clientelism. *Annual Review of Political Science, 14*, 289–310.

Hilgers, T. (2008). Causes and consequences of political clientelism: Mexico's PRD in comparative perspective. *Latin American Politics and Society, 50*(4), 123–153.

Hilgers, T. (2011). Clientelism and conceptual stretching: Differentiating among concepts and among analytical levels. *Theory and Society, 40*(5), 567–588.

Instituto Nacional de Estadística y Geografía (INEGI) (2010). Censo de Población y Vivienda. Available at: www.beta.inegi.org.mx/proyectos/ccpv/2010/ [Accessed on 30 October 2016].

International Energy Agency (IEA) (2016). Mexico Energy Outlook: World Energy Outlook Special Report. Paris, France.

Jin, Y., Ma, X., Chen, X., Cheng, Y., Baris, E., and Ezzati, M. (2006). Exposure to indoor air pollution from household energy use in rural China: The interactions of technology, behavior, and knowledge in health risk management. *Social Science and Medicine, 62*(12), 3161–3176.

Kyu, H. H., Georgiades, K., and Boyle, M. H. (2010). Biofuel smoke and child anemia in 29 developing countries: A multilevel analysis. *Annals of Epidemiology, 20*(11), 811–817.

Lewis, J. J. and Pattanayak, S. K. (2012). Who adopts improved fuels and cookstoves? A systematic review. *Environmental Health Perspectives, 120*(5), 637.

Lim, S. S., Vos, T., Flaxman, A. D., Danaei, G., Shibuya, K., Adair-Rohani, H. … and Aryee, M. (2012). A comparative risk assessment of burden of disease and injury attributable to 67 risk factors and risk factor clusters in 21 regions, 1990–2010: A systematic analysis for the Global Burden of Disease Study 2010. *The Lancet, 380*(9859), 2224–2260.

López-Martínez, O. and Cuanalo de la Cerda, H. E. (2017). Participatory action research in the design, construction and evaluation of improved cook stoves in a rural Yucatec Maya community. *Action Research*, 0(0), 1-20.

Lucas, A., (2016). Elite capture and corruption in two villages in Bengkulu Province, Sumatra. *Human Ecology, 44*(3), 287–300.

Lukes, S. ([1974] 2002). *Power: a radical view*.in: Haugaard, M., ed. (2002) *Power: A Reader*, Manchester: Manchester University Press.

Masera, O., Edwards, R., Arnez, C. A., Berrueta, V., Johnson, M., Bracho, L. R. … and Smith, K. R. (2007). Impact of Patsari improved cookstoves on indoor air quality in Michoacán, Mexico. *Energy for Sustainable Development, 11*(2), 45–56.

Modi, V., McDade, S., Lallement, D., and Saghir, J. (2005). *Energy Services for the Millennium Development Goals*. (UN Millennium Project.) New York: UNDP, World Bank and ESMAP.

Muneer, S. E. T. (2003). Adoption of biomass improved cookstoves in a patriarchal society: An example from Sudan. *Science of the Total Environment, 307*(1), 259–266.

Nepal, M., Nepal, A., and Grimsrud, K. (2011). Unbelievable but improved cookstoves are not helpful in reducing firewood demand in Nepal. *Environment and Development Economics, 16*(01), 1–23.

Olvera, A. (2016). Political crisis, social movements and the future of democracy in Mexico. *Revista Mexicana de Ciencias Politicas y Sociales, 61*(226), 279–295.

Osei, R. D. (2010). Toyola charcoal stove: Improving the environment and health of the poor in Ghana. GIM Case Study (B095).

Parikh, J. K. (1995). Gender issues in energy policy. *Energy Policy, 23*(9), 745–754.

Person, B., Loo, J. D., Owuor, M., Ogange, L., Jefferds, M. E. D., and Cohen, A. L. (2012). "It is good for my family's health and cooks food in a way that my heart loves": Qualitative findings and implications for scaling up an improved cookstove project in rural Kenya. *International Journal of Environmental Research and Public Health, 9*(5), 1566–1580.

Rambaldi, G., Kyem, P. A. K., McCall, M., and Weiner, D. (2006). Participatory spatial information management and communication in developing countries. *The Electronic Journal of Information Systems in Developing Countries, 25*.

Rotmans, J. and Loorbach, D. (2010). Towards a better understanding of transitions and their governance. A systemic and reflexive approach. *Transitions to Sustainable Development. New Directions in the Study of Long Term Transformative Change*. Washington DC: World Bank, 105–198.

Ruiz-Mercado, I., Masera, O., Zamora, H., and Smith, K. R. (2011). Adoption and sustained use of improved cookstoves. *Energy Policy, 39*(12), 7557–7566.

Sacramento Rivero, J. C., Eastmond-Spencer, A., Becerril García, J., and Navarro-Pineda, F. S. (2016). A three-dimensional sustainability evaluation of Jatropha plantations in Yucatan, Mexico. *Sustainability, 8*(12), 1316.

Scott, J. (2001). *Power*. Cambridge: Polity.

Sidanius, J. and Pratto, F. (2001). *Social Dominance: An Intergroup Theory of Social Hierarchy and Oppression*. Cambridge University Press.

Sidanius, J. and Pratto, F. (2011). Social dominance theory. *Handbook of Theories of Social Psychology, 2*, 418–438.

Smith, K. R., Shuhua, G., Kun, H., and Daxiong, Q. (1993). One hundred million improved cookstoves in China: How was it done? *World Development, 21*(6), 941–961.

Standing, A. and Hilson, G. (2013). Distributing mining wealth to communities in Ghana: Addressing problems of elite capture and political corruption. *U4 Issue, 2013*(5).

Troncoso, K., Castillo, A., Masera, O., and Merino, L. (2007). Social perceptions about a technological innovation for fuelwood cooking: Case study in rural Mexico. *Energy Policy, 35*(5), 2799–2810.

USAID/Winrock (2009). Commercialization of Improved Cookstoves for Reduced Indoor Air Pollution in Urban Slums of Northwest Bangladesh. Washington, DC: United States Agency for International Development. Available at: pdf.usaid.gov/pdf_docs/pnado851.pdf [Accessed on 2 August 2011].

Venkataraman, C., Sagar, A. D., Habib, G., Lam, N., and Smith, K. R. (2010). The Indian national initiative for advanced biomass cookstoves: The benefits of clean combustion. *Energy for Sustainable Development, 14*(2), 63–72.

PART IV
Resilience

10

THE ROLE OF VOLUNTARY AGREEMENTS IN A HYBRID MODEL OF ENVIRONMENTAL LAW ENFORCEMENT

Joice Chang

Traditional approaches to environmental regulation have largely been deterrence-based: laws are established to discourage regulated entities from noncompliance, and they have long been criticized for various reasons, including economic inefficiencies, lack of flexibility, difficulty with enforcement, and low compliance. As such, many scholars, politicians, and regulators have looked to either supplementary or even alternative policy instruments to solve environmental problems, and more specifically, improve compliance (see Hughes, 2012). Among the myriad possibilities, one that has gained traction as an attractive alternative in recent years involves voluntary agreements (VAs). Broadly, VAs are "agreements between government and industry to facilitate voluntary action with a desirable social outcome, which is encouraged by the government, to be undertaken by the participant based on the participant's self-interest" (Storey, Boyd, and Dowd, 1999, p. 188). With a few exceptions, VAs are used mostly in environmental regulatory realms (one of the notable exceptions involved various voluntary agreements between the tobacco industry and the United Kingdom government, such as one to restrict tobacco advertising near schools, see Bryden et al., 2013).

In reality, the elements of particular VAs differ greatly, but these approaches share a commitment to voluntary cooperation. There is a substantial and growing body of literature that supports the view that a VA is more efficient and effective than a deterrence-based approach; it is viewed as a cooperative approach that minimizes the negative political and legal consequences of regulatory failure. However, some scholars are more cautious, suggesting that VAs undermine more stringent and powerful regulatory tools (eg, Lyon and Maxwell, 2004) and that VAs should be viewed as a limited policy instrument more useful in settings where more powerful ones are infeasible.

The next section of this chapter introduces the main differences between deterrence-based and cooperation-based approaches in terms of motivations for compliance, strategies for enforcement, and methods of enforcement. Even though cooperation-based approaches, such as VAs, are gaining traction, the overall enforcement scheme in environmental regulation remains mainly deterrence-based. The following section argues that, as a result, regulated entities and regulators operate in a hybrid environment. Drawing from and extending existing social science and legal scholarship on compliance and enforcement, I propose a framework for conceptualizing the role of VAs in this environment. Then, applying the framework to several hypothetical scenarios, I examine the implications of VAs in environmental regulation for compliance and regulatory behaviors. Based on that analysis, the chapter concludes with a discussion of the importance of understanding issues of social context, scale, and resilience when considering environmental policy instruments.

Differences between deterrence-based and cooperation-based approaches

In general, approaches to environmental law enforcement can be broken down into two types: deterrence-based and cooperation-based (Coffee, 1992; Garvie and Keeler, 1994; Firestone, 2002). Deterrence-based approaches are the traditional strategy for regulating noncompliance. The assumption underlying deterrence-based approaches is that decision makers at a regulated entity (hereinafter, "decision makers") are rational economic actors who aim to maximize profits. Thus, the decision to comply or not with environmental policy depends on whether the costs of noncompliance outweigh its benefits. The task for the regulator, then, is to set both the probability of detection and the penalties high enough that it becomes irrational for the firm to not comply. This model of enforcement focuses on the detection of noncompliance with the law.

In contrast, cooperation-based approaches to enforcement are premised on different assumptions about behavior; decision makers are assumed to be motivated by a mix of factors, and, on the whole, are inclined to comply with the law. Further, cooperation-based approaches posit that regulators should treat those they regulate like partners. In other words, if the latter are found to be noncompliant, they may respond positively to suggestions and advice on how to achieve compliance. Unlike the focus in deterrence-based approaches on detection of noncompliance, cooperation-based models of enforcement seek to influence the process of compliance.

Based on these divergent assumptions about the motivations for behavior among decision makers, the two approaches to enforcement also offer different conceptualizations of the strategies regulators use to enforce laws and the effects of enforcement, as summarized in Table 10.1. The remainder of this section will discuss these two models in detail.

Motivations: calculated vs. normative/social

The two approaches to environmental enforcement differ in how they perceive the regulated firms' motivations for complying. Because deterrence-based models

TABLE 10.1 Deterrence-based enforcement vs. cooperation-based enforcement

Enforcement Considerations	*Enforcement Approaches*	
	Deterrence-Based	*Cooperation-Based*
Motivations for compliance	Calculated	Normative/social
Strategies of enforcement	Formalistic	Facilitative
Primary method of enforcement	Punishment or compensation	Deterrence
Secondary method of enforcement	Deterrence	Punishment or compensation

assume economic welfare-maximization motivations among rational actors, decision makers are also assumed to engage in the calculation of the benefits and costs of compliance and to subsequently choose the option (compliance or noncompliance) that yields the higher net return. The fact that these calculations are presumed to take place and shape compliance behaviors matters in how enforcement happens. Under this approach, the likelihood of detection and the characteristics (eg, speed, certainty, and magnitude) of the punishment imposed both influence the expected response of any given regulated firm. Thus, compliance rates are expected to be greater for firms who believe that there is a stronger likelihood of detection as well as an economic penalty imposed for noncompliance (Helland, 1998; Stafford, 2002; 2006; Gray and Shadbegian, 2007). In addition, because the calculus of expected utility draws attention to the costs of compliance, the rate of compliance is expected to be lower as the costs of compliance increases.

In contrast, cooperation-based approaches recognize that motivations for compliance may be more complex. In particular, one of the reasons firms comply is their internalized normative commitment or the sense that they have a duty to comply with the law (Winter and May, 2001). In general, compliance is greater when regulated entities have a stronger sense of normative commitment to compliance (Winter and May, 2001). Cooperation-based approaches also account for social motivation among decision makers to earn the respect of the relevant parties with whom they regularly interact (Grasmick and Bursik, 1990). The sources for such pressures may include other regulated firms, regulators, citizen groups, and the media. The desire to comply may, over time, socialize the decision makers to normatively commit to compliance. However, social commitments are distinctly different from normative commitments in that the former may lead a decision maker at a firm to comply in response to values (eg, approval of others) that may not have been internalized.

Strategies of enforcement: formalistic/deterrence-based vs. facilitative/cooperation-based

For both models of enforcement, regulatory enforcement is viewed as a game that involves, over time, multiple interactions – both formal and informal – between the regulator and the decision makers (see Ayres and Braithwaite, 1992). These repeated interactions lead to a set of shared expectations for both the regulator and the decision makers. The formal aspects of such repeated interactions set forth rules and consequences of noncompliance, while the informal aspects establish social relationships between the regulators and those they regulate. However, the models of enforcement differ vastly in how they conceptualize the management of interactions via inspection strategies.

The primary focus of deterrence-based approaches is on using formalistic enforcement strategies and adhering to legalistic interpretation and application of the rules to enhance the regulated entity's compliance. Formalism can be seen in strict standards, implementation deadlines, and levels of penalties (see Hawkins, 1984; Kagan, 1994). The main arguments for formalistic enforcement strategies are frequently couched in terms of consistency, as less administrative discretion may prevent corruption, and in terms of equity and fairness (Bardach and Kagan, 1982; Gunningham, 1987; Langbein, 2002). However, formalistic approaches may lead the decision makers to view the rules as unreasonable, which may subsequently engender a hostile relationship between them and the regulator.

A formalistic regulator is aware that the decision makers must have knowledge of the relevant rules and, equally importantly, must have the capacity to comply with said rules. Compliance capacity can be viewed as a combination of knowledge, experience, and a working relationship with regulators (May and Wood, 2003). While the regulator may be cognizant of the decision makers' ignorance of the law or inability to comply, s/he is likely not going to take that into consideration when applying the rules. After all, when noncompliance is detected, deterrence-based approaches focus on the outcome and the legal sanctions for that violation. The propensity is to impose, or at the very least signal the imposition of, sanctions for noncompliance. Because the deterrence-based model of enforcement does not incorporate an educational function to increase awareness or understanding of the law, knowledge of the law and capacity to comply do not matter much when sanctions for noncompliance are imposed.

In contrast, cooperation-based approaches stress the value of a cooperative relationship between the regulator and the decision makers. This scheme uses facilitative strategies, and the two sides are not viewed as adversaries, but as partners. As such, the regulator's main focus is on assisting the decision makers as they work to achieve compliance in a cooperative climate (Sparrow, 2000). In such an environment, the regulator expects the decision makers to do their best to comply, and in turn, the decision makers expect the regulator to not be too rigid in applying the rules of enforcement. A major advantage of this approach is its explicit recognition of the iterative nature of the regulator–regulated entity relationship. However, the drawback of an enforcement regime that values flexibility and cooperation is that it may result in inconsistent and/or unfair applications of the rules (May and Wood, 2003).

One of the facilitative regulator's primary aims is to enhance the decision makers' knowledge of the rules and the attendant capacity to comply. The emphasis on facilitation is underpinned by the belief that noncompliance may be due to ignorance of the law, lack of ability to comply (or incompetence), and/or negligence. Thus, the regulator in a cooperation-based model of enforcement facilitates compliance by using more flexible guidelines in compliance assessment, explaining to the noncompliant decision makers the regulations violated and advising how to fix them, using incentives such as fewer inspections or even leniency when noncompliance is detected with decision makers who demonstrate good faith effort to comply, and providing technical assistance to the decision makers (McKay, 2003). The emphasis on flexible or selective enforcement takes into consideration the particular circumstances under which decision makers committed a violation (Glicksman and Earnhart, 2007), allowing for a contextualized approach to regulatory compliance that takes knowledge and existing capacity into account and places responsibility on the regulator for facilitating both knowledge and capacity to achieve compliance.

Methods of enforcement

In general, there are two main methods of environmental enforcement: deterrence and punishment (of the wrongdoer) or compensation (for the victims). Each approach has a primary and secondary effect in enforcement (see Figure 10.1). Regulators following deterrence-based approaches take a variety of administrative or judicial actions in response to noncompliance. Inspections are designed to detect violations and to collect evidence to build a case against an offending firm. The imposition of penalties or the initiation of other enforcement action is viewed as a mark of success, and such sanctions are expected to deter future noncompliance (Arbuckle, 1993). However, because noncompliant decision makers are not penalized until after the violation has been detected, the main effect of this model of enforcement is actually *ex post* punishment or compensation. The use of *ex post* punishment induces future compliance. In other words, the intended effect

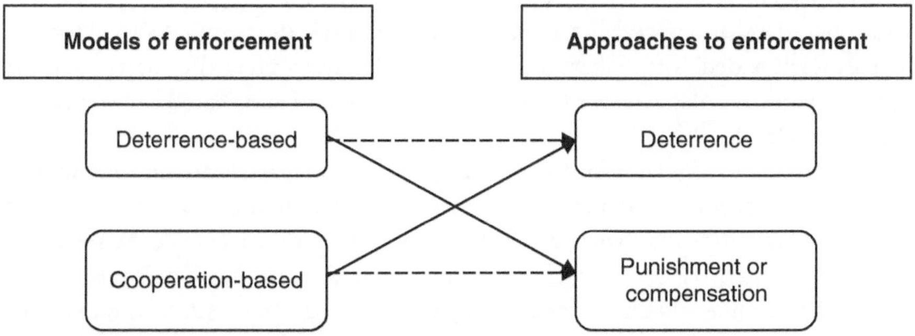

FIGURE 10.1 Legal and regulatory context of environmental enforcement
Source: Joice Chang

of deterrence becomes secondary, and the imposition of penalties or compensation becomes the primary approach to enforcement.

In contrast, under cooperation-based approaches, inspections are meant to provide advice to firms about compliance. In addition, regulators and decision makers engage in informal bargaining rather than adversarial procedures to resolve noncompliance. Instead of penalties, regulators prefer using rewards and incentives to achieve compliance. In fact, the imposition of penalties and initiation of formal actions are viewed as the system's inability to obtain compliance. Thus, cooperation-based approaches primarily focus on securing compliance rather than punishment (see Rechtschaffen, 2000). That is, its main effect is *ex ante* deterrence, and secondarily, its residual effect is punishment or compensation.

Reality: a hybrid model of enforcement

While there are signs that cooperation-based approaches are gaining traction in recent years, environmental enforcement remains largely deterrence-based; thus, decision makers and regulators interact in a hybrid environment (see Glicksman and Earnhart, 2007). The increasing use of VAs represents a move toward more cooperation. By participating in VAs, regulators and decision makers are sending signals of their respective commitments to a more cooperative enforcement environment.

In a hybrid regime of enforcement, calculated motivations matter, but so do normative and social motivations (Glicksman and Earnhart, 2007). Both actual inspections and the threat of detection contribute to increased compliance (see Harrington, 1998). At the same time, normative and social motivations are just as influential as calculated motivations in bringing about compliance (Grasmick and Bursik, 1990; Winter and May, 2001). In addition, a regulator utilizes both formalistic and facilitative strategies. The use of formalism firmly establishes the rules as well as the consequences of noncompliance. In doing so, the regulator contributes to deterrence that is specific to decision makers at a particular regulated firm. Concurrently, the same regulator's interactions with the decision makers are iterative and adaptive, and they may establish a less adversarial relationship. The regulator uses facilitative strategies to achieve and maintain a regulatory partnership characterized by more collaboration and mutual trust (Sparrow, 2000). Through the partnership, decision makers are then more likely to view the interpretation/application of the rules as a set of expectations negotiated and shared between them and the regulator.

Aside from the regulator and decision makers, other parties are also relevant in the hybrid enforcement scheme. Environmental enforcement is more complex than a series of interactions between the regulator and decision makers at the regulated firm. Judicial and quasi-judicial actors are involved in the resolution of issues of noncompliance. Unlike judicial actors, quasi-judicial actors (e.g., Administrative Law Judges, or ALJs) are in the executive branch; they are judicial in nature since they are responsible for the administration of justice. Judicial and quasi-judicial actors will be referred to as "judicial actors" hereinafter. An examination of environmental enforcement not only must recognize its "hybrid" nature, it also must explicitly identify the pertinence of judicial actors.

Role of VAs in a hybrid model of enforcement

Given the context discussed above, what is the role of VAs in a hybrid model of enforcement? On the one hand, through the use of regulatory incentives and the provision of technical assistance, a VA allows regulators and decision makers to engage in more cooperative relationships. On the other hand, the implementation of such instruments does not negate the legal aspects of environmental law enforcement. In fact, many VAs actually demand better compliance records from their participants (than from non-participating counterparts). "Compliance" here refers to adherence to existing legal requirements that are applicable to all firms, and not to the requirements of a VA. Depending on the program, compliance with VAs can be self-reported or independently assessed. Next, VA-participating firms still face inspections (for compliance to legal requirements), and when violations are detected, they remain subject to penalties. Thus, the use of strategies from both deterrence-based and cooperation-based approaches accounts for and taps into all three motivations for firm compliance. Because enforcement does not cease when a VA is introduced, decision makers continue to partially base decisions on the expected costs and benefits of compliance. At the same time, the "voluntary" aspect of a VA means that the decision makers choose to enter into such a partnership because they are also socially or normatively committed to compliance; it also establishes the regulator and decision makers (which are still subject to existing legal requirements) as partners.

Further, how does the presence of judicial actors affect the role of VAs in hybrid environmental enforcement? In a strictly deterrence-based setting, regulators expect decision makers to respond rationally to the costs and benefits of compliance. Judicial actors' decisions regarding existence of or penalty for noncompliance would be made within the boundaries of the existing law. Since VAs are not legally binding, the fact that a regulated entity is a VA participant would not, in theory, be relevant in how penalties are decided. However, in a hybrid setting, the judicial actor, similar to the regulator, may also view decision makers as partners. To judicial actors, VA membership may signal social or normative commitment to compliance. The violation detected, then, could be viewed as an unusual aberration that can be explained and possibly punished less severely.

The above framework, which explicitly recognizes the hybrid nature of environmental enforcement as well as the relevance of judicial actors, should be used to examine the effects of VAs on the outcomes of environmental regulation. To properly understand the role of VAs in enforcement, it is necessary to examine how VA participation affects conditions both external and internal to a firm (see Borck, Coglianese, and Nash, 2006). External conditions can be regulatory, economic, or social. Internal conditions include managerial incentives, organizational identity, organizational culture, organizational commitments, managers' personal affiliations,

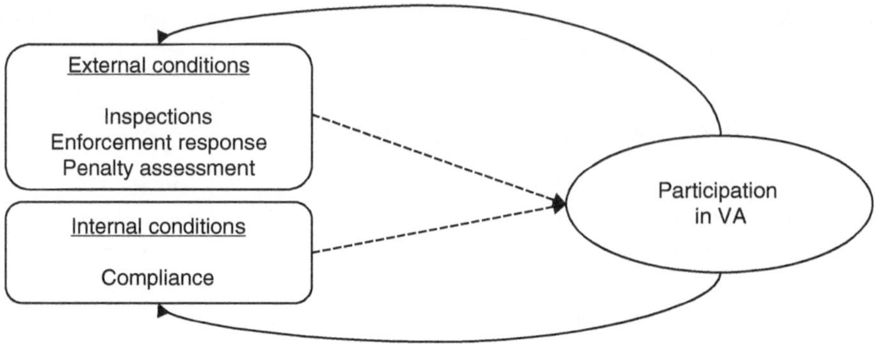

FIGURE 10.2 Bi-directional relationship between VA participation and compliance and enforcement

and relationships with external stakeholders. While conditions can be variable and wide-ranging, the most relevant ones in this particular context are compliance behaviors (internal condition) and regulatory behaviors (external condition). Whether or how a regulated entity complies with the law is a proxy for its organizational decision-making calculus (an internal condition). Regulators respond with varying levels of inspections and penalties (an external condition). Much of the literature on VAs has focused on the effects of both of these conditions on the decision to participate in VAs (see King and Lenox, 2000; Videras and Alberini, 2000; Khanna et al., 2007; Innes and Sam, 2008). However, the reverse relationship is equally important; that is, the decision to participate (or not), after it is made, can also influence the same set of conditions. Thus, the bi-directional nature of the relationship between VA participation and compliance behavior, as well as regulatory behaviors in the form of inspections, enforcement responses, and penalty assessments, must be recognized (see Figure 10.2).

Implications

In the hybrid model of environmental enforcement in which a VA exists, what is the nature of the relationship between decision makers, regulators, and judicial actors? In other words, from the perspective of decision makers, how do internal and external conditions change (if at all) in response to the addition of a variable – namely, a VA? To begin answering these questions, the above conceptual framework presents a complex picture of VAs and environmental enforcement, as well as some useful implications for analyzing environmental policy instruments.

Broadly, an empirical analysis of VAs could be based on a principal–agent model. More specifically, a screening model could be assumed (Mas-Colell, Whinston and Green, 1995). In such a model, the principal would be the regulators and the agents would be the decision makers at regulated VA-participating firms, and the contract to which they agree would be the VA. In this situation, information asymmetry on noncompliance exists: while the agents have information on their own violations, the principal cannot possibly detect all of them. In other words, the regulators are imperfectly informed about the characteristics of the participating firms (see Rasmusen, 1994; Salanié, 2005). Then, the informed players (decision makers) respond after the program has been formulated (ie, after the uninformed player has acted). Figure 10. 3 provides a visualization of the chain of decisions in the relationship between the principal (regulator) and the agents (decision makers).

Implications for compliance behaviors

Literature on why decision makers comply in general and on their motivations for joining VAs in particular is extensive (see King and Lenox, 2000; Videras and Alberini, 2000; Khanna et al., 2007; Innes and Sam, 2008). However, there is a distinct disconnect between these two areas of research. The positive theoretical research on compliance suggests that there are various reasons for firm compliance, including the desire to improve or maintain a certain corporate image (see Harford, 1997). In the VA literature, there is a seemingly disparate set of reasons for why firms participate, and three potential motivations for joining VAs are consistent with the aforementioned reason for compliance. Firms often decide to participate in VAs because they want to attract "green consumers" (Videras and Alberini, 2000; Khanna et al., 2007), they are responding to stakeholder pressures (King and Lenox, 2000), and/or they wish to atone for past poorer performance records (King and Lenox, 2000; Videras and Alberini, 2000). These reasons, then, are all variants of the decision makers' concern with the image the firm presents to consumers, surrounding communities, and/or regulators.

The research is mixed on the question of what is the relationship, if any, between VA participation and compliance. Decision makers can enhance their firm's image by participating in a VA, which may act as a proxy for good behavior (see King and Lenox, 2000; Videras and Alberini, 2000). Such an act increases the firm's level of voluntary investment as an observable sign of commitment to further improve compliance, although the actual impact is ambiguous (see Lyon and Maxwell, 2004). On the one hand, the investment would reduce the marginal cost of compliance effort, and therefore would increase effort, all else equal. On the other hand, because the probability of compliance increases, the regulatory effort, *ceteris paribus*, would decrease. This, then, would indirectly reduce the firm's cost of noncompliance, providing an incentive for decision makers to reduce compliance effort (Lyon and Maxwell, 2004).

However, the particular design of any VA might matter; the more rigorous the structure of a VA (eg, application processes, monitoring, and sanctions), the higher the costs of participation. VA firms, when compared to similar non-VA firms, are more likely to be connected with and be visible to surrounding communities, consumers, environmental groups, and regulators (Borck, Coglianese, and Nash, 2006). Thus, all else equal, the cost of participation and the level of exposure to the public would be positively correlated with higher *ex post* compliance rates for VA participants.

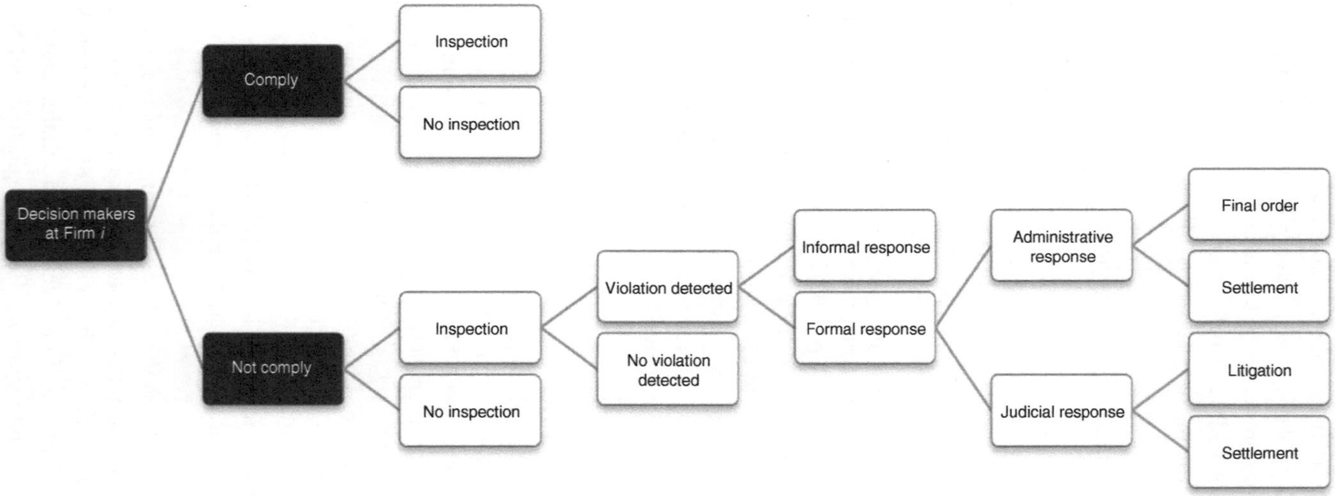

FIGURE 10.3 Chain of decisions

Source: Joice Chang

Implications for regulatory behaviors

There are various theories on how regulators enforce environmental laws, ranging from their focus on specific deterrence (Firestone, 2002), to the emphasis on general deterrence (Coffee, 1992; Garvie and Keeler, 1994), to their relationship to political and community pressures (Helland, 1998). However, one idea stands out as particularly useful for understanding what those monitoring and enforcing compliance actually do: responsive regulation. In brief, responsive regulation is the notion that in order to be effective, regulations need to take into account and be able to respond to the variety of motivations, characteristics, and needs of the regulated community (Ayres and Braithwaite, 1992; Decker, 2007). More specifically, then, regulators enforce legal requirements responsively (Decker, 2007). In this model, regulators are flexible and take into account firm-specific characteristics, such as membership in a VA, when deciding when to inspect and when determining responses to detected noncompliance. However, responsive enforcement must be earned; that is, decision makers must provide evidence of good behavior, such as improved compliance. Therefore, this model recognizes that while enforcement takes place in a mostly deterrence-based environment, repeated interactions between regulators and the regulated community lead both sides to use strategies to signal a willingness to cooperate.

Assume a world with two firms, *F1* and *F2*, as well as a regulator, *R*. Here, *R* is, generally, the agency responsible for environmental law enforcement, like the EPA; therefore, *R* could be an inspector, an attorney, or a judge that works for the agency. Empirical research on inspections has shown that the probability of *R* inspecting the firms depends on their respective prior compliance records and emission levels; firms with prior noncompliance tend to receive more inspections and firms with higher emissions are inspected more often (Kleit, Pierce, and Hill, 1998). In a purely deterrence-based setting, instruments that foster cooperation between regulator and decision makers, such as VAs, would not exist. If similar in all respects, both *F1* and *F2* would face the same probability of being inspected. Further, the primary goal of a deterrence-based approach is punishment after the violation has taken place and been detected. Therefore, *R* would heavily favor more "formal" enforcement mechanisms or those that are implemented to punish (rather than to deter future noncompliance). If *F1* and *F2* were to be found in noncompliance for similar infractions, one should expect not only that the enforcement actions they face would be similar, but also that they would face more formal actions, such as administrative and judicial proceedings.

More realistically, however, the regulatory setting in the United States and other countries is a mixture of deterrence-based and cooperation-based approaches. In this scenario, VAs exist. Assume that *F1* is a VA participant and *F2* is not, but both are similar in all other respects. Under these circumstances, decision makers are sending different signals to *R*; by committing to overcompliance, *F1* is able to show *R* that it is, at least in some ways, a higher-quality firm than *F2*. If such an observable commitment is made prior to the *R*'s allocation of inspection resources, *R* will likely respond with fewer future inspections of *F1* than of *F2*.

Similarly, such a signal of good behavior may influence *R* to be more cooperative in penalty assessment and enforcement actions. Because the main concern of cooperation-based approaches is deterrence of future violations, *R* would have to demonstrate, in response to *F1*'s credible commitment to overcompliance (eg, joining a VA), its own credible commitment to increased cooperation. As such, *R* would be more interested in maintaining a non-adversarial relationship with *F1* than with *F2*. *R* would likely choose to engage in more informal bargaining (ie, by issuing non-binding letters and notices) in the event of detected noncompliance. For similar offenses, the likelihood that *R* would take informal rather than formal actions against a firm would be higher for *F1* than for *F2*, since *R* would be responding positively to *F1*'s signal of good behavior via VA participation.

In a cooperative regulatory setting, penalties are typically either reduced or disfavored altogether. Instead of imposing penalties, *R* would prefer using rewards and incentives to achieve compliance. Even so, *R* would not completely forego formal sanctions; there would be violations that *R* would find necessary to penalize. For these cases, *R* would view VA participation positively, which may be helpful in civil and administrative proceedings (Hemphill, 1993/94). Therefore, in cases in which penalties are assessed, they would likely be lower for *F1* than for *F2*.

Furthermore, the adjudicatory environment in which formal enforcement actions are resolved matters. Most such actions are resolved administratively (see Rosenberg, 2008). Most enforcement actions initiated by the EPA involve assessment of civil penalties and generally deal with less egregious violations. In contrast, enforcement actions brought by the Department of Justice are litigated in federal district courts and involve more serious violations. Moreover, there are no specific rubrics for calculating final civil penalties (see Rosenberg, 2008). This may leave regulators, including ALJs and civil court judges, with room to adjust the penalties according to their preferences. In an administrative case, the ALJ generally defers to the EPA's recommendation on the issue of penalty. In contrast, the same kind of deference is not as explicit in civil cases. Therefore, one should expect that not only would penalties be generally lower when decided administratively, they would also be lower for VA participants, like *F1*.

Last, the manner in which a case is resolved is also critical. Once *R* decides to bring a case forward to an adjudicatory body – administrative or judicial – it can be resolved through either litigation or settlement. In general, *R* would favor settling disputes (Rosenberg, 2008). In litigation (whether administrative or judicial), *R* would likely seek higher penalties than in settlement (Mugdan, 2000). Because settling a dispute is consistent with the cooperation-based approaches, *R* would more likely engage both *F1* and *F2* in settlements. However, in similar cases, the penalties resulting from a settlement with *F1* would be lower than that with *F2*, reflecting *R*'s responsiveness to *F1*'s signal of superior behavior.

Conclusion

Environmental policy instruments, regardless of whether they are deterrence-based or cooperation-based, can advance or hinder possibilities for improved environmental outcomes, including compliance and regulatory behaviors. In the US and other countries, regulators and regulated entities operate in a hybrid environment, which is largely deterrence-based, with the recent injection of cooperation-based approaches. In particular, this chapter examines how VAs – a mostly cooperation-based approach – can alter compliance and regulatory behaviors in this hybrid environment. The analysis of the role of VAs highlights the importance of several issues when considering environmental policy instruments.

First, there are key internal and external conditions that affect whether and how decision makers choose to participate in VAs. Internally, decision makers may decide to voluntarily join a VA because they are normatively and socially committed to compliance. Externally, regulators may be eager to enter into such an arrangement because they recognize the iterative nature of the regulator–regulated entity relationship and are aiming to facilitate improved future compliance behaviors (rather than to punish past noncompliance). The reverse relationship – how VA participation affects internal and external conditions – is equally important for understanding environmental policymaking. Here, the internal and external conditions of interest are compliance and regulatory behaviors, respectively. VAs, in theory, should lead to improved compliance because they facilitate more cooperation between regulators and regulated entities. However, this analysis suggests that this may not be so straightforward; the cost of participation in the VA and the level of the firm's exposure to the public are factors that may complicate whether or how VA participation affects compliance behaviors. Furthermore, VA participation affects regulatory behaviors. When a firm joins a VA, it sends a signal of good behavior to regulators. This, in turn, has cascading effects in inspections and enforcement actions. Compared to those at an otherwise similar but non-VA firm, decision makers at a VA-participating firm send a signal of superior performance (they are, after all, taking on more obligations than legally required) to the regulators. Because this is an observable commitment made prior to the allocation of inspection resources, regulators may likely respond with fewer inspections of the latter firm. Likewise, the same signal may lead regulators to respond with less punitive actions and penalties in order to maintain a more cooperative relationship with the VA-participating regulated entity.

The issue of scale is similarly crucial when considering environmental policy instruments. A large-scale, one-size-fits-all (eg, deterrence-based) approach to environmental regulation is prone to various regulatory failures, such as low compliance rates and economic inefficiencies. Instead, alternative policy instruments, such as VAs, could facilitate a more interactive relationship between regulators and regulated entities. VAs, in particular, allow for decision makers to send distinguishing signals to regulators. This, in turn, provides regulators the ability to take into account firm-specific characteristics, such as VA membership, when making enforcement decisions.

The last factor that is critical to thinking about environmental policy instruments is resilience, in terms of both the long-term success of the firm and the long-term effectiveness of the policy instrument. Decision makers join VAs mainly because they are concerned with the image the firm presents to its stakeholders, such as consumers, shareholders, community members, and regulators. In other words, they recognize that environmental compliance may be positively correlated to their economic success in the long run. As a policy instrument, VAs allow regulators to consider firm-specific characteristics in the enforcement of environmental regulations. This kind of flexibility may engender a less adversarial and thus a potentially more long-lasting cooperative relationship in the regulation of environmental enforcement.

References

Arbuckle, J. G. (1993). Water pollution control. In: J. G. Arbuckle, ed., *Environmental Law Handbook*, 12th ed. Rockville, MD: Government Institute, pp. 151–220.

Ayres, I. and Braithwaite, J. (1992). *Responsive Regulation: Transcending the Deregulation Debate.* New York, NY: Oxford University Press.

Bardach, E. and Kagan, R. A. (1982). *Going by the Book: The Problem of Regulatory Unreasonableness.* Philadelphia, PA: Temple University Press.

Borck, J., Coglianese, C., and Nash, J. (2006). Why do they join? An exploration of business participation in voluntary environmental programs. In: *Beyond Compliance: Business Decision Making and the U.S. EPA's Performance Track Program*, 1st ed. Cambridge, MA: John F. Kennedy School of Government, Harvard University.

Bryden, A., Petticrew, M, Mays, N., Eastmure, E., and Knai, C. (2013). Voluntary agreements between government and business: a scoping review of the literature with specific reference to the Public Health Responsibility Deal. *Health Policy*, 110(2–3), pp. 186–197.

Coffee, J. C., Jr. (1992). Paradigms lost: the blurring of the criminal and civil law models. And what could be done about it. *Yale Law Journal*, 30(1), pp. 1875–1893.

Decker, C. S. (2007). Flexible enforcement and fine adjustment. *Regulation and Governance*, 1(4), pp. 312–328.

Firestone, J. (2002). Agency governance and enforcement: the influence of mission on environmental decisionmaking. *Journal of Policy Analysis and Management*, 21(3), pp. 409–426.

Garvie, D. and Keeler, A. (1994). Incomplete enforcement with endogenous regulatory choice. *Journal of Public Economics*, 55(1), pp. 141–162.

Glicksman, R. L. and Earnhart, D. H. (2007). The comparative effectiveness of government interventions on environmental performance in the chemical industry. *Stanford Environmental Law Journal*, 26, pp. 317–371.

Grasmick, H. G. and Bursik, R. J., Jr. (1990). Conscience, significant others, and rational choice: extending the deterrence model. *Law and Society Review*, 24(3), pp. 837–861.

Gray, W. B. and Shadbegian, R. J. (2007). The environmental performance of polluting plants: a spatial analysis. *Journal of Regional Science*, 47(1), pp. 63–84.

Gunningham, N. (1987). Negotiated non-compliance: a case study of regulatory failure. *Law and Policy*, 9(69), pp. 69–95.

Harford, J. D. (1997). Firm ownership patterns and motives for voluntary pollution control. *Managerial and Decision Economics*, 18(6), pp. 421–431.

Harrington, W. (1998). Enforcement leverage when penalties are restricted. *Journal of Public Economics*, 37(1), pp. 29–53.

Hawkins, K. (1984). *Environment and Enforcement: Regulation and the Social Definition of Pollution*. New York, NY: Clarendon Press.

Helland, E. (1998). The enforcement of pollution control laws: inspections, violations, and self-reporting. *Review of Economics and Statistics*, 80(1), pp. 141–153.

Hemphill, T. A. (1993/94). Corporate environmentalism and self-regulation: keeping enforcement agencies at bay. *Journal of Environmental Regulation*, 3(20), pp. 145–154.

Hughes, S. (2012). Voluntary environmental programs in the public sector: evaluating an urban water conservation program in California. *Policy Studies Journal*, 40(4), pp. 650–673.

Innes, R. and Sam, A. G. (2008). Voluntary pollution reductions and the enforcement of environmental law: an empirical study of the 35/50 Program. *Journal of Law and Economics*, 51(2), pp. 271–296.

Kagan, R. A. (1994). Do lawyers cause adversarial legalism? A preliminary inquiry. *Law and Social Inquiry*, 19(1), pp. 1–62.

Khanna, M., Koss, P., Jones, C., and Ervin, D. (2007). Motivations for voluntary environmental management. *Policy Studies Journal*, 35(4), pp. 751–772.

King, A. A. and Lenox, M. J. (2000). Industry self-regulation without sanctions: the chemical industry's Responsible Care Program. *The Academy of Management Journal*, 43(4), pp. 698–716.

Kleit, A. N., Pierce, M. A. and Hill, R. C. (1998). Environmental protection, agency motivations, and rent extraction: the regulation of water pollution in Louisiana. *Journal of Regulatory Economics*, 13(2), pp. 121–137.

Langbein, L. I. (2002). Responsive bureaus, equity, and regulatory negotiation: an empirical view. *Journal of Policy Analysis and Management*, 21(3), pp. 449–465.

Lyon, T. P. and Maxwell, J. W. (2004). *Corporate Environmentalism and Public Policy*. New York, NY: Cambridge University Press.

Mas-Colell, A., Whinston, M. D., and Green, J. R. (1995). *Microeconomic Theory*. New York: Oxford University Press, pp. 436–467.

May, P. J. and Wood, R. S. (2003). At the regulatory front lines: inspectors' enforcement styles and regulatory compliance. *Journal of Public Administration Research and Theory*, 13(2), pp. 117–139.

McKay, S. (2003). Sheriffs and outlaws: in pursuit of effective enforcement. *Town Planning Review*, 74(4), pp. 423–443.

Mugdan, W. E. (2000). Federal environmental enforcement in EPA Region 2. *Albany Law Environmental Outlook Journal*, 5(1), pp. 14–20.

Rasmusen, E. (1994). *Games and Information: An Introduction to Game Theory*. 2nd ed. Cambridge, MA: B. Blackwell.

Rechtschaffen, C. (2000). Competing visions: EPA and the states battle for the future of environmental enforcement. *Environmental Law Reporter*, 30(10), pp. 10803–10828.

Rosenberg, R. H. (2008). Doing more or doing less for the environment: Shedding light on EPA's "stealth" method of environmental enforcement. William and Mary Law School Faculty Publications, Paper 94.

Salanié, B. (2005). *The Economics of Contracts: A Primer*. 2nd ed. Cambridge, MA: MIT Press.

Sparrow, M. K. (2000). *The Regulatory Craft: Controlling Risks, Solving Problems, and Managing Compliance*. Washington, DC: Brookings.

Stafford, S. L. (2002). The effect of punishment on firm compliance with hazardous waste regulations. *Journal of Environmental Economics and Management*, 44(2), pp. 290–308.

Stafford, S. L. (2006). State adoption of environmental audit initiatives. *Contemporary Economic Policy*, 24(1), pp. 172–187.
Storey, M., Boyd, G., and Dowd, J. (1999). Voluntary agreements with industry. In: C. Carraro and F. Lévêque, eds., *Voluntary Approaches in Environmental Policy*, 1st ed. Boston: Kluwer Academic Publishers, pp. 187–207.
Videras, J. and Alberini, A. (2000). The appeal of voluntary environmental programs: which firms participate and why? *Contemporary Economic Policy*, 18(4), pp. 449–460.
Winter, S. C. and May, P. J. (2001). Motivation for compliance with environmental regulations. *Journal of Policy Analysis and Management*, 20(4), pp. 675–698.

11

INSTITUTIONAL BARRIERS TO MANAGING DYNAMIC LANDSCAPES

Lessons learned in Southwestern Colorado

Nina Burkardt and Patricia Biddle Orth

Managing in the face of climate change creates many challenges for public land and water managers as they seek to implement practices that support the sustainability of both natural and human systems. Managers are required to protect resources and to support allowable uses and activities, but determining the balance between protection and use depends on policy directives, agency mandates, and analysis of present and predicted future conditions. A variety of federal laws and policies direct decision makers to consider climate effects, but the uncertainty associated with understanding climate trajectories makes policy creation and implementation difficult.

In many areas of the western United States, residents and public land managers are experiencing the effects of climate change. Studies have documented warmer temperatures and altered water availability in this region and the effects of those events on agriculture, wildfire occurrence, and human health (EPA, 2016). In this chapter, we describe a project that included a multi-disciplinary group of natural and social scientists and a diverse group of stakeholders in the San Juan Basin in southwestern Colorado. The intent of the project was to assist land managers to develop climate adaptation strategies to promote social and ecological resiliency and sustainability. Determining effective policy approaches for sustainable land and water management practices is an ongoing challenge for federal lands, and climate change adds a layer of complexity to this effort. The project team developed a Social-Ecological Adaptation Planning Framework for this work, using components adapted from Stein et al., 2014; Cross et al., 2012; and Murphy et al., 2016 (Figure 11.1).

The Upper San Juan Basin in southwestern Colorado is an important example of the challenges of sustainability and the interplay of climate change, public land uses, and human communities. The basin covers nearly four million acres, with 45% in federal land ownership, 21% in tribal ownership, 33% in private ownership, and the remainder in state, city, or local hands (Colorado Natural Heritage Program, n.d.). These lands provide a variety of recreational benefits, access for

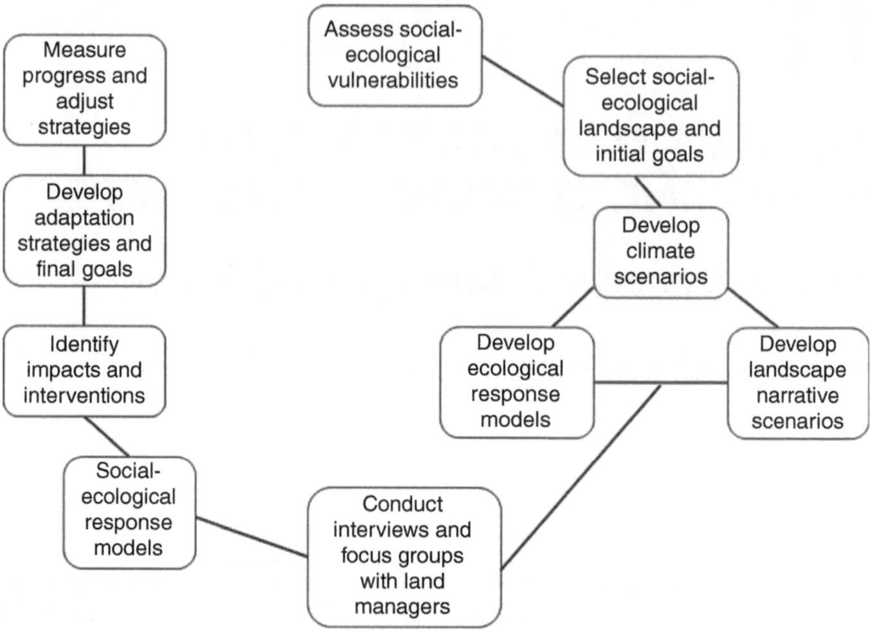

FIGURE 11.1 Social-ecological adaptation planning framework developed and applied through this project

Source: Adapted from Stein et al., 2014; Cross et al., 2012; and Murphy et al., 2016

grazing and other uses, and a host of ecosystem services to enhance human well being, including provisioning, regulating, supporting, and cultural services (Millennium Ecosystem Assessment, 2005). Climate change may affect the ability of public lands to provide these benefits. For example, if ecosystems are stressed because of rising temperatures and altered patterns of precipitation, ranchers may be compelled to change grazing patterns, recreational uses may be curtailed, or water deliveries may not be made.

An important component of the project was developing an understanding of institutional and policy barriers to the creation and implementation of climate adaptation strategies. Much has been written on this subject in recent years, and some of the barriers noted in the literature include institutional, attitudinal, economic, and political factors (Eckstrom and Moser, 2012); unclear mandates, agency cultures that hinder adaptation, and insufficient stakeholder support (Archie et al., 2012; Jantarasami et al., 2010); lack of management-relevant climate science, inconsistent agency guidance, and lack of resources (Kemp et al., 2015; Eisenack et al., 2014; Archie et al., 2012). Others note that the Endangered Species Act (ESA; 16 U.S.C. § 1531 et seq.) and other prescriptive laws may hinder adaptation because of their focus on single species and lack of flexibility (Jantarasami et al., 2010). In some cases, requirements of the National Environmental Policy Act (NEPA) are seen as supportive of adaptation, both because NEPA focuses on procedural requirements and because it requires analysis of more than one management alternative, which potentially provides room for climate adaptation actions (Jantarasami et al., 2010). In this chapter, we describe a method for working with stakeholders used in the San Juan Basin to develop adaptation strategies, and discuss perceptions of barriers to developing and implementing those strategies. Although stakeholders acknowledged the necessity of planning for adaptation, many discussions revolved around the institutional and policy barriers to actually developing and implementing adaptation strategies. Understanding barriers, where they originate, how they affect adaptation, and how they can be overcome are important questions for anyone with a role in promoting policies for social-ecological resilience and sustainability in the face of a changing climate.

Background and study approach

Over the course of the project, the team worked with local stakeholders on a frequent basis to increase the likelihood that the findings and outcomes would be useful and would promote the development of adaptation strategies. The project focused on developing a linked ecological-climate-social framework to depict the landscape inhabited by the stakeholders, which required ongoing conversations and discussion of findings as they were produced. This overview of the process and products aims to explain how the group arrived at the discussions about barriers to adaptation. While the activities built on one another, in some cases they were conducted simultaneously.

Action 1: select landscapes or adaptation targets

One project objective was to focus on specific landscape features or attributes that were meaningful to stakeholders and decision makers in their locations. The project team convened meetings to coproduce criteria for selection of these landscapes and then worked with attendees to apply the criteria and select adaptation targets. These targets became the basis for all of the following project outputs.

Action 2: develop climate scenarios

A climate modeler used two emissions scenarios (Representative Concentration Pathway (RCP) 4.5 and RCP 8.5) and 72 climate models to show a range of projections for two climate variables, temperature and precipitation, for the period from 2015 to 2035. Three climate scenario clusters emerged as a result of this analysis. They were labeled "Hot and Dry," "Warm and Wet," and "Feast or Famine." No judgment was made as to the likelihood of occurrence of one scenario over another; according to the models, all are plausible.

Action 3: conduct interviews

Social scientists conducted 34 in-depth, semi-structured interviews with stakeholders who live and work in the basin. The purpose of the interviews was to learn about the use, importance, and status of the adaptation targets, how changes to those targets would affect human communities, and how stakeholders approach dealing with uncertainty. Interviews were recorded, transcribed, and analyzed using a content-analysis software package.

Action 4: develop ecological response models

Ecological response models were developed to assess how each adaptation target would fare under the three climate scenarios (Rondeau et al., 2017). For example, for pinyon-juniper landscapes in the San Juan Basin, the models showed that a longer and more intense fire season in the "Hot and Dry" scenario was likely to result in less regeneration of pinyon-juniper and a transition to a shrub-dominated system. Insects and disease outbreaks under this scenario are likely to increase tree mortality, leading to changes in forest composition and impacts to species that rely on pinyon-juniper ecosystems.

Action 5: develop narrative scenarios

The climate models and ecological response models were used to write narrative scenarios that described climate and climate impacts for the local area for each of the three climate scenarios. Similar to the climate models, all of the narrative scenarios were considered to be plausible, but no probabilities were assigned to the likelihood of their occurrence. The purpose of the scenarios was to translate ecological and climate information into stories to allow stakeholders to glimpse potential futures and consider how to respond.

Action 6: conduct focus groups and workshops to respond to narrative scenarios

Stakeholders were presented with the narrative scenarios and asked to identify how they thought they would respond to the impacts. While every scenario produced concern, "Feast or Famine" was seen as the most problematic because of the potential extremes of precipitation and temperature and the difficulty of planning for these events. Nonetheless, this process helped stakeholders to see the connections between high-level climate science projections and local impacts. It also illustrated the effects of uncertainty on the process of planning for climate impacts: how can long-term decisions effectively account for climate change impacts when uncertainty is high?

Action 7: develop social-ecological models

Two different facilitated participatory modeling exercises were conducted with stakeholders for each adaptation target. Although the models were somewhat different in their operations, they both allowed stakeholders to apply an "If-Then" approach to the effects of each of the climate scenarios and to identify cascading effects. As the models were developed, participants had the opportunity to assign a quantitative level of scientific uncertainty (1–4, with 1 being highly uncertain and 4 being highly certain, for example) and identify points of intervention. This allowed stakeholders to see the connections between social and ecological effects and to think about the interplay of climate impacts.

Action 8: develop strategies and actions

This was a complex process of determining specific objectives for the adaptation targets and then identifying strategies that could achieve those objectives at a landscape level and across more than one climate scenario. As an example, in one workshop, the three adaptation strategies for pinyon-juniper were to: 1) identify and protect target species by developing refugia, which are locations where species are expected to persist; 2) assist and allow transformation, recognizing that ecosystem change may be inevitable; and 3) conduct proactive treatment to enhance forest resilience.

Action 9: identify barriers to developing and implementing adaptation strategies

Identification of barriers to adaptation occurred in many stages of the project, but most deliberately during interviews and strategy development workshops. The interview transcripts were one source of information for a barriers assessment, and records from the strategy workshops were also evaluated. Participants were well aware that taking action required overcoming a number of obstacles, and these were discussed openly. Managers and those in leadership positions outside the local area were also contacted by phone and e-mail to determine the consistency of barrier identification across levels of an organization.

Action 10: evaluate strategies, develop monitoring and evaluation plan (in progress)

At the conclusion of the project, investigators acknowledged that monitoring the deployment of adaptation strategies was a necessary next step. Questions remain about how to discern the effects of the barriers, whether the strategies had the expected effects, and whether climate change paths are more or less uncertain over time. Because the objective of the strategies is to promote sustainable ecosystems and human communities, one anticipated outcome for stakeholders is promoting changes to policy to support adaptation. Developing a better understanding of how strategies are implemented will help in the evaluation of how current policies provide support for adaptation and whether policy changes are necessary to strengthen adaptive capacity.

Analysis of barriers

We used several data sources to categorize and analyze perceived barriers to adaptation. Thirty-four semi-structured interviews of federal and state agency personnel and other stakeholders were transcribed and analyzed using content analysis software. We employed a grounded theory approach to coding (Strauss and Corbin, 1990), in which we allowed themes to emerge from the data rather than using only a set of pre-determined codes. The social science researcher who conducted the interviews developed the initial coding scheme, but in that analysis, the primary focus was not on learning about barriers to adaptation. Nonetheless, interviewees offered thoughts about barriers to adaptation and these were noted in the first round of coding (Wyborn et al., 2015). A second round of coding focused specifically on barriers and yielded additional coded interview segments.

We also interviewed seven non-local managers and agency leaders across the state level, regional offices, and headquarters offices in Washington, DC. Individuals with responsibility for developing or implementing climate change policies were targeted for these conversations. We wanted to learn about their definitions of climate adaptation, perceptions of barriers to climate change adaptation and how they thought they could be overcome, and how they gave direction to others in the agency about climate change adaptation. We were particularly interested in the final question because at the field level many mentioned a lack of guidance on climate adaptation from higher organizational levels as a specific adaptation barrier. Five interviews were completed on the telephone, and two managers elected to respond to our questions in written format. We analyzed the content of these seven conversations using the themes that were developed when we coded interview transcripts.

Another source of information for our barriers analysis was records of the final workshops in which stakeholders articulated potential specific adaptation actions. A facilitated discussion included thoughts about barriers to the actions that participants had worked hard to develop during the workshop. As with the narrative scenario workshops, these items were presented in a written report (Rondeau et al., 2017), and we included them in our analysis.

Findings

Our analysis revealed five broad categories of barriers to developing and implementing adaptation strategies: uncertainty about climate change impacts; agency capacity; decision-making processes; knowledge barriers; and public skepticism about the need for climate adaptation.

Uncertainty about climate change impacts

Although several of the identified barriers are inter-dependent, the barrier of uncertainty is a recurring theme. Most participants in the interviews and workshops felt challenged to deal with climate impacts due to uncertainty, and it is possible that more certainty about the trajectory of the climate would make many of the other barriers less important. Uncertainty for our participants was related to the nature, extent, and magnitude of climate impacts, because not knowing those things creates a highly challenging decision environment.

Nonetheless, public land managers are accustomed to making decisions in an uncertain environment. Management plans are developed to function over ten-year periods or longer and to accommodate disturbances on the landscape such as fires or beetle infestations; periods of drought; extreme weather events; and political, social, and cultural shifts. The question for many participants was whether these past reference points and changes were adequate for understanding and planning for the future. At the same time, planning and management activities are ongoing and dynamic and are not halted by uncertainty.

The people we interviewed had a variety of reactions to uncertainty, from a pragmatic attitude and realization that they deal with uncertainty all the time, to dread and even a sense of paralysis. For the former group, thinking of climate effects as a disturbance fits existing frameworks for land management and some are hopeful that existing land management practices allow them sufficient flexibility to manage for climate change. Others said they found it daunting to think about the aggressive treatment that would be required to respond to climate impacts and were less confident of the adequacy of existing policies and practices. Uncertainty seemed to lead some to consider how they would pursue risk-averse strategies and make incremental changes in management, lacking confidence in their ability to predict the future. Others noted that uncertainty produced a fear of litigation and expressed the need for defensible data to reduce that concern. This fear, some noted, would be likely to reduce flexibility, experimentation, and other conditions required for adaptation. Finally, there were discussions of how uncertainty affected the level of trust of the public in the ability of land managers to make decisions that reflected climate concerns. This was seen as especially problematic in addressing climate scenarios that show high inter-annual climate variability with no clear long-term directional trends

Agency capacity

Agency culture is the reflection of shared values in a group and is marked by signs or symbols (level of formality among members, status markers such as office space, etc.), espoused beliefs and behavioral norms, and underlying assumptions, which may be taken for granted and are often largely unseen (Schein, 2004). When describing barriers related to agency capacity, interviewees pointed to the structural and cultural internal elements of their organizations. Many described how insufficient budgets and staffing stymied efforts to plan for climate adaptation or to monitor ecological change, which was seen as increasingly important in times of uncertainty. Bureaucratic factors suggest a culture that is slow to respond to change, structured to be reactive rather than proactive, and risk averse. These inclinations may be exacerbated by fragmented land ownership patterns and the tendency of agencies to focus on short-term activities rather than long-term ecological change. For example, in the pinyon-juniper landscape, managers are more likely to manage for specific uses such as grazing, recreation, and timber management than to view the landscape as a whole.

An important component of agency capacity and culture is leadership, and a consistent theme across the field-level interviews and workshops was a lack of clear guidance on how to manage for climate adaptation. In our discussions with regional and national agency leadership – those to whom field-level personnel look for direction – individuals noted a need for leadership from those even higher in their organizations and some indicated that without that direction it was difficult to know how to reply to queries from the field.

Additional input on barriers to adaptation came from workshops to develop on-the-ground adaptation actions. Workshop attendees mentioned many of the barriers already discussed; in addition, they raised the challenge of insufficient funding to implement actions and continue long-term management. To those at the field level, such practical aspects of climate adaptation may be especially relevant.

Decision-making processes

Public land managers operate under specific policies, planning documents, and decision processes that are designed to comply with agency mandates and provide some level of certainty for resources and land users. National Forests develop Forest Management Plans that consider forest activities and their effects for periods of ten years or more (Janowiak et al., 2014; Wells, 2014). BLM offices create Resource Management Plans, and National Parks operate under Park Management Plans. In addition, some activities on federally managed lands require review under the National Environmental Policy Act (NEPA; 42 U.S.C. § 4321 et seq.). Some study participants noted that none of these planning or regulatory processes account for change or uncertainty because they are based on the assumption of stationarity – in the case of climate, the notion that past climate conditions and climate responses are indicators of the future. Further, planning and regulatory processes are often slow, sometimes taking several years for scientific studies, public input, and agency approval. While management plans have some flexibility and allow for response to discrete events, planning for adaptation was thought to stretch the ability of many land management agencies. Some participants questioned how well uncertainty is incorporated into NEPA analysis and how responsive the NEPA process can be to climate factors in the long term. Another challenge for decision makers is that federal budget cycles run in one-year increments while long-term planning for adaptation may be a multi-decades endeavor. Aligning decisions with budget cycles was viewed as a barrier to adaptation planning.

As noted previously, the workshops for adaptation strategy development enabled participants to consider specific actions, rather than adaptation more broadly, and in this setting, we also heard comments regarding the inflexibility of existing policies that may constrain the adoption of some actions. As examples, some noted that agency policies might not allow planting seed from outside the region, which was one suggested adaption action. Others noted that fire management goals and policies may need to be revised to implement adaptation strategies.

Knowledge barriers

Interviewees expressed concerns about three knowledge barriers related to climate information. First, there is a lack of knowledge about climate change, which some referred to as a shortage of climate literacy in the agencies. Agency professionals are trained in forestry, biology, ecology, and the like, and use those disciplinary skills in their work. Climate literacy may be an additional skill set to enhance professional competency, but may not be provided in the workplace. The second knowledge-related barrier is a lack of expertise within agencies to interpret climate change models and information. Lacking this expertise, agency personnel are dependent on relationships with climate scientists or others who are able to perform a translation function, as someone with specific expertise is needed to help managers and decision makers apply this information.

The third knowledge barrier is lack of access to credible, scale-relevant data to inform planning and decision-making. Some of the study participants noted that they make decisions at relatively small scales – a National Forest, for example – and need peer-reviewed data to help them understand climate effects and trajectories at those scales. Both local and national level respondents noted the need to receive climate information at an appropriate scale and level of specificity for managers trying to understand resources and make decisions. Even in cases when science is conducted, making the connection between science and policy is often not well done.

During the adaptation strategy workshops, participants discussed the problems of a lack of knowledge about the impacts of climate change on ecosystems and low confidence in predicting which areas are most likely to be affected. Because of the specificity of the strategy workshops, other knowledge barriers focused on scientific topics in need of additional research for each strategy, such as weighing the benefits of removing juniper and pinyon versus the weed problems this causes, and understanding the effects of accepting pinyon-juniper invasion into the emergent zones.

Public skepticism about the need for climate adaptation

Policies and decisions to promote climate adaptation on public lands may require public support in order to be successful, especially if those decisions affect users of those lands. For example, allowing an ecosystem to transform from sagebrush to pinyon-juniper may result in a less culturally valued landscape, and some members of the public may prefer that management decisions focus on maintaining existing species rather than assisting transformation. In the interviews, managers noted public skepticism as an important barrier to adaptation.

In the adaptation strategy workshops, attendees clearly articulated their understanding that adaptation actions required public acceptance or support. As the groups outlined adaptation actions to achieve those strategic objectives, they noted barriers. Those related to public support included public understanding, acceptance, and buy-in on the need for refugia, lack of public awareness regarding climate change, and perceptions that cultural preferences and values of desired ecosystems may not align with transitioning ecosystems on the ground. While some barriers regarding public acceptance are based on skepticism of climate change, others are grounded in lack of public understanding of climate effects or a lack of support for actions that agencies might take to manage for adaptation.

Discussion and conclusions

Our work in the San Juan Basin helped us to understand better how public land managers approach climate adaptation, how they perceive barriers to adaptation, and what policy options are seen as promising approaches for supporting sustainable social-ecological systems. Managers in this study felt challenged to deal with climate adaptation due to uncertainty about how climate effects will play out on a scale that is relevant to their decisions (Rondeau et al., 2017). This was compounded by a lack of high-level guidance, incremental and risk-averse agency decision processes, inflexible regulatory processes, lack of capacity to respond, and lack of public support for adaptation actions. Taken together, these barriers suggest a challenging policy environment in which current decision processes may be inadequate, but substantial policy change is also unlikely. Existing legislation and agency policies, including NEPA and agency-specific planning processes, may provide the strongest frameworks for climate adaptation. Policy innovations that address the identified barriers could provide additional flexibility in existing policies and support decision processes that offer increased opportunities for inclusion of interested stakeholders.

One policy approach to address many of the barriers to the implementation of adaptation strategies is to promote and support adaptive management. In adaptive management, resource management is an iterative learning process and actions are evaluated in terms of their efficacy in attaining desired conditions. Uncertainty is acknowledged, results are monitored, and policies or practices are adjusted as outcomes are discerned (Williams, Szaro, and Shapiro, 2009). Adopting a policy paradigm of adaptive management can be one way to make decisions in situations of uncertainty and can insert some flexibility into agency decisions. As with other policy approaches to address barriers to adaptation planning, adaptive management requires leadership, commitment, resources, and willingness of managers and others to adjust their actions over time to achieve objectives.

A second policy approach to overcome barriers to adaptation is to strengthen and support stakeholder engagement processes. The case reviewed in this chapter was not part of a formal decision process but laid the foundation for development of adaptation strategies; implementing those strategies may involve a more structured setting. Stakeholders and managers spent almost three years selecting adaptation targets, digesting the science about how climate change could affect those targets, and discussing potential ways to adapt. Thus, relationships were built that are likely to be important in taking the next steps.

Another potential benefit of ongoing and robust engagement is that it can address the challenge of designing and implementing landscape-scale adaptation actions in areas of fragmented management. The San Juan Basin includes land in tribal ownership, lands managed by multiple federal and state agencies, and privately held land. Determining adaptation actions over this ownership mix is challenging, and no single entity has the capacity or authority to carry out this task. A high level of cooperation and coordination will be required to overcome this barrier, and working together, as did these San Juan Basin stakeholders, may support development of relationships that are key to cooperation.

A third approach for overcoming barriers is coproduction of knowledge. Coproduction means that scientists and science users work together to produce usable information. This requires time to build relationships and develop communication channels that allow scientists and stakeholders to understand what is needed for decision-making, and what can realistically be produced (Meadow et al., 2015; Dilling and Lemos, 2011). In this project, the study team established relationships with resource managers and other stakeholders before initiation of the project and worked to strengthen those relationships and produce usable knowledge. Coproduction could be a useful model as managers move forward with establishing climate literacy, targeting science needs for developing adaptation strategies, and communicating across jurisdictional boundaries to develop landscape-scale solutions. Creating policies to encourage coproduction requires the support of leadership and incentives such as funding, time, and staff to participate in these activities.

Development of local capacity to respond to change is also an important policy consideration. Adaptation actions are very particular to place and scale, and what is effective in one landscape may not be in another. Differences in biophysical, social, and institutional factors should be accounted for when considering successful policy implementation and adaptation actions. Policies that provide high-level guidance and support while leaving details of implementation to regional or local actors may be best suited to address issues of scale and differing social-ecological systems.

A final policy consideration is building conflict resolution and trust-building skills within agencies. Some participants noted that future conditions were likely to create new types of conflict among resource users because of changing resource availability, inadequate mechanisms to deal with ecosystem changes, and shifting of the distribution of costs and benefits in an altered landscape. Because landscape-scale adaptation will require coordination among numerous entities, agency conflict resolution and trust-building skills are likely to become increasingly important. Agencies across the study area have differing mandates and objectives, and building cooperative mechanisms may become increasingly important if climate effects become more pronounced.

The process described in this chapter may support sustainable social-ecological systems by helping affected parties develop a shared view of current conditions and potential trajectories of change. Building on a joint understanding, land managers and other stakeholders can develop strategies to sustain human and natural systems in the face of uncertainty. This requires investment in scientific expertise and a willingness to experiment with promising policy approaches to developing and implementing adaptation strategies. As communities across the United States undertake climate adaptation actions, it is essential to monitor the results of adaptation decisions to determine which approaches are most effective, where adaptation barriers exist, and how innovative policies can support sustainability in a changing climate.
Acknowledgements: this research was funded by the Department of the Interior's North Central Climate Science Center and the USDA Forest Service Rocky Mountain Research Station.

References

Archie, K. M., Dilling, L., Milford, J. B., and Pampel, F. C. (2012). Climate change and western public lands: a survey of U.S. federal land managers on the status of adaptation efforts. *Ecology and Society*, [online] 17(4), p. 20. Available at: www.ecologyandsociety.org/vol17/iss4/art20/ [Accessed 16 June 2016].

Colorado Natural Heritage Program (n.d.). *Wetland Profile for the Upper San Juan Basin* [online]. Fort Collins, CO. Available at: www.cnhp.colostate.edu/wetlandinventory/profile.asp?GeoUnitID=10 [Accessed 1 May 2017].

Cross, M. S., Zavaleta, E. S., Bachelet, D., Brooks, M. L., Enquist, C. A. F., Fleishman, E., Graumlich, L. J., Groves, C. R., Hannah, L., Hansen, L., Hayward, G., Koopman, M., Lawler, J. L., Malcom, J., Nordgren, J., Petersen, B., Rowland, E. L., Scott, D., Shafer, S. L., Shaw, M. R., and Tabor, G. M. (2012). The Adaptation for Conservation Targets (ACT) framework: a tool for incorporating climate change into natural resource management. *Environmental Management,* 50, pp. 341–351.

Dilling, L. and Lemos, M. C. (2011). Creating usable science: opportunities and constraints for climate knowledge use and their implications for science policy. *Global Environmental Change*, 21, pp. 680–689.

Eckstrom, J. A. and Moser, S. C. (2012). Identifying and overcoming barriers in urban climate adaptation: case study findings from the San Francisco Bay Area, California, USA. *Urban Climate*, [online] 9, pp. 54–74. Available at: www.sciencedirect.com/science/article/pii/S221209551400042X [Accessed 10 February 2016].

Eisenack, K., Moser, S. C., Hoffmann, E., Klein, R. J. T., Oberlack, C., Pechan, A., Rotter, M., and Termeer, C. J. A. M. (2014). Explaining and overcoming barriers to climate change adaptation. *Nature Climate Change*, 4 (10), pp. 867–872.

Environmental Protection Agency (EPA) (2016). *What Climate Change Means for Colorado*. Washington, D.C: EPA 430-F-16-008 [online]. Available at: 19january2017snapshot.epa.gov/sites/production/files/2016-09/documents/climate-change-co.pdf [Accessed 18 July 2017].

Janowiak, M. K., Swanston, C. W., Nagel, L. M., Brandt, L. A., Butler, P. R., Handler, S. D., Shannon P. D., Iverson, L. R., Matthews, S. N., Prasad, A., and Peters, M. P. (2014). A practical approach for translating climate change adaptation principles into forest management actions. *Journal of Forestry*, 112(5), pp. 424–433.

Jantarasami, L. C., Lawler, J. J., and Thomas, C. W. (2010). Institutional barriers to climate change adaptation in U.S. National Parks and Forests. *Ecology and Society*, [online] 15(4), p. 33. Available at: www.ecologyandsociety.org/vol15/iss4/ [Accessed 4 November 2014].

Kemp, K. B., Blades, J. J., Klos, P. Z., Hall, T. E., Force, J. E., Morgan, P., and Tinkham, W. T. (2015). Managing for climate change on federal lands of the western United States: perceived usefulness of climate science, effectiveness of adaptation strategies, and barriers to implementation. *Ecology and Society* [online] 20(2), p. 17. Available at: www.ecologyandsociety.org/vol20/iss2/ [Accessed 12 June 2017].

Meadow, A. M., Ferguson, D. B., Guido, Z., Horangic, A., Owen, G., and Wall, T. (2015). Moving toward the deliberate coproduction of climate science knowledge. *Weather, Climate, and Society*, 7, pp. 179–191.

Millennium Ecosystem Assessment (2005). *Ecosystems and Human Well-being: Synthesis* [online]. Washington, DC: Island Press, p. 155. Available at: www.millenniumassessment.org/documents/document.356.aspx.pdf [Accessed 16 October 2009].

Murphy, D. J., Wyborn, C., Yung, L., Cleveland, C., Eby, L., Dobrowski, S., Towler, E., and Williams, D. R. (2016). Engaging communities and climate change futures with multi-scale iterative scenario-building (MISB) in the Western United States. *Human Organization*, 75(1), pp. 33–46.

Rondeau, R., Bidwell, M., Neely, B., Rangwala, I., Yung, L., and Wyborn, C. (2017). *Pinyon-Juniper Landscape: San Juan Basin, Colorado* [online]. Ft. Collins, Colorado: North Central Climate Science Center, p. 39.

Schein, E. H. (2004). *Organizational Culture and Leadership*. 3rd edition. San Francisco: Jossey-Bass.

Stein, B. A., Glick, P., Edelson, N., and Staudt, A. (eds.) (2014). *Climate-Smart Conservation: Putting Adaptation Principles into Practice* [online]. Washington, DC: National Wildlife Federation. Available at: www.nwf.org/pdf/Climate-Smart-Conservation/NWF-Climate-Smart-Conservation_5-08-14.pdf [Accessed 24 October 2016].

Strauss, A. and Corbin, J. (1990). *Basics of Qualitative Research*, 15. Newbury Park, CA: Sage.

Wells, K. (2014). Can't see the trees for the forest? The ongoing controversy over assessing the site specific impacts of Comprehensive Forest Management Plans. *Ecology Law Quarterly*, 41(2), pp. 553–580.

Williams, B. K., Szaro, R. C., and Shapiro, C. D. (2009). *Adaptive Management: the U.S. Department of the Interior Technical Guide*. Washington, DC: US Department of the Interior, Adaptive Management Working Group.

Wyborn, C., Yung, L., Bidwell, M., and Rondeau, R. (2015). Climate adaptation in the San Juan Basin: interviews on pinyon-juniper and seeps, springs, and wetlands [online]. Available at: static1.squarespace.com/static/53bc5871e4b095b6a42949b4/t/56b281972eeb817ddb25b776/1454539159990/Climate+Adaptation+Interview+Report+FINAL.pdf [Accessed 10 November 2015].

12

THE LONG ROAD TO SUSTAINABILITY OF MEXICAN OIL PALM PRODUCTION

Erin C. Pischke

Environmental policies often outline myriad goals and numerous ways to achieve them. For example, policies may dictate a change in land use or crop production while promoting social development for communities. Yet environmental policies often fall short of meeting their goals. Federal policies that promote the expansion of oil palm in Mexico are examples of this type of measure that falls short of meeting its social goals as the crop expands and is used to meet other goals, such as diversifying commodity crops. There are few comprehensive policies that establish how oil palm production will be grown in an explicitly sustainable manner in terms of promoting the well being of society, the long-term profitability of the crop, and protection for the health of the environment (Solomon et al., 2015; Stattman and Mol, 2014). In the absence of strong federal policies, third-party certification and international sustainability standards offer the ability to meet such development goals while also promoting economically viable and environmentally safe oil palm production.

Since the 1990s, oil palm has been grown in the Mexican countryside (Diario Oficial de la Federación, 1998). The crop can be used to produce many diverse products (eg, soap, cosmetics, frying oil) but can only grow in tropical regions at certain latitudes (Byerlee, Falcon, and Naylor, 2017). Since oil palm demand is growing for cooking oil derived from oil palm seeds in countries like China and India, feedstock growth in countries that produce it is expanding (Byerlee, Falcon, and Naylor, 2017). In Mexico, the crop has also been identified as one among many that has the ability to fulfill policy mandates for creating biofuels while also diversifying the agricultural sector (Eastmond et al., 2014; Rodríguez, Vázquez, and Gamboa, 2014).

Oil palm production is a common candidate for sustainability certification offered by private third parties. Certification of oil palm globally has grown since the largest certifying organization's inception in 2004 (Byerlee, Falcon, and Naylor,

2017; RSPO, 2017a). International sustainability standards are becoming more commonplace, as consumers and investors demand corporate social responsibility from global companies (Byerlee, Falcon, and Naylor, 2017).

This chapter explores the linkages between sustainability of oil palm production in Mexico and the opportunity that voluntary third-party certification and international sustainability standards have to improve socio-economic and environmental conditions in local communities where the crop is grown. The chapter first reviews the scale of production of oil palm in Mexico and the related policies that the Mexican government has enacted, before presenting the certification and international sustainability standards that the country has agreed to follow. Then I present a case study of Tabasco, Mexico, from work on a National Science Foundation-funded project, Partnerships in International Research and Education. The focus of the project was to address the socio-ecological sustainability of bioenergy development across six countries in the Americas (Argentina, Brazil, Canada, Mexico, the United States, and Uruguay), including the sustainability of oil palm production in Tabasco. The discussion section covers the challenges and deficiencies the country faces in meeting the sustainability criteria as well as opportunities for sustainable oil palm production in the country's future. Despite the lack of government oversight or concern for the sustainable production of oil palm and the potential benefits third-party governance would procure through certification in terms of promoting sustainable production of the crop, I will show that the experience with producing oil palm in Mexico in general, and Tabasco in particular, has a long way to go before it can truly offer a means of sustainable economic and environmental resilience for local communities, national economies, and global markets.

Oil palm production in Mexico

Mexico is the world's seventeenth largest producer of oil palm, producing around 140,000 metric tons of oil palm per year (USDA-FAS, 2017) in the states of Campeche, Chiapas, Tabasco, and Veracruz (Rubio, 2017). There are 119,000 hectares of productive oil palm planted in the country, with 4,287 hectares of it grown in Tabasco State (Rubio, 2017). As of 2012, the Secretary of Agriculture, Livestock, Rural Development, Fisheries and Food (SAGARPA) had identified a potential of over two million hectares of land suitable for planting oil palm in Mexico (Gobierno de Mexico, 2012), with 324,000 hectares available for cultivation in Tabasco (SEDAFOP, 2017). The Secretariat of Agricultural Development, Forestry and Fisheries (SEDAFOP) has the ability to provide 150 oil palm producers in the state with 1,500 hectares of plants (SEDAFOP, 2017). Thus, the country has room to expand its production and the means to support producers to do so.

Policies promoting oil palm expansion

Mexico has a centralized federal government, with policies being determined and coordinated by the federal government in a top-down manner, dictating the creation of state policies, which determine what happens at the local level (Romero-Hernández et al., 2011). Through environmental policies such as the 2008 Bioenergy Law and commitments to international treaties, Mexico has often promoted various agricultural crops and commodities as a way to ensure rural development while diversifying the energy sector and reducing the country's greenhouse gas (GHG) emissions (Cámara de Diputados, 2008; Diario Oficial de la Federación, 2008). The National Development Plan for the period 2013–2018 outlines ambitions to increase productivity and promote "green jobs" in the country through investment, employment opportunities, and diversifying the agricultural sector, which includes subsidizing biofuel feedstocks like oil palm (Gobierno de la Republica, 2013).

There are also two energy sector policies – 2008 Law for the Development of Renewable Energy and Energy Transition Financing (LAERFTE) and the 2012 Mexican General Law for Climate Change – that are indirectly involved in promoting oil palm production through bioenergy development (Alemán-Nava et al., 2014; Solomon et al., 2015; Rodríguez, Vázquez, and Gamboa, 2014). Most recently, the country has pledged to mitigate climate change through agricultural, forestry, and land reforms, the introduction of a carbon tax, enactment of energy efficiency measures, and increased development of alternative fuels such as biodiesel, which can be produced from oil palm (UNFCCC, 2016). These policies have social and economic components to them that prioritize production yields over the well being of communities where they are implemented, and few have environmental protections for biodiversity hot spots, demonstrating a need for policies that promote sustainability of socio-ecological systems across the integrated issues of environmental protection, economic resilience, and community well being.

Regardless of policy makers' willingness to pass legislation that promotes sustainability, problems with weak governance and a lack of resources can hinder a country's ability to implement such policies (Jordana, 2010). Furthermore, it has been challenging for developing countries to meet goals outlined in international treaties and federal environmental policies (Selin and VanDeveer, 2009). The Mexican government can influence which programs the states can receive money for, but does little to help finance or coordinate implementation (Valenzuela, 2014). In addition to limited funding and coordination efforts, subnational governments like Tabasco's do not always have the technical expertise or human capacity necessary to meet their goals (Gay and Estrada, n.d.).

Third-party certification and international sustainability initiatives

Typical governmental control is structured by a top-down, command-and-control approach to policymaking by a central state actor (Tollefson, Gale, and Haley, 2008). However, a national leadership may lose control of their means to govern because of the effects of globalization and democratization (Howlett,

2000). In this condition, non- or quasi-governmental organizations can take on the roles of government (Hooghe and Marks, 2003; Howlett, 2000; Howlett, Ramesh, and Perl, 2009). In the context of global concern for environmental problems like climate change, multi-stakeholder initiatives created by international stakeholders (eg, NGOs, environmental organizations, corporations, and global finance organizations) play a role in governing environmental management in various ways. These entities, often driven by their own interests, strive to address issues such as rural poverty through programs that manage GHG emissions (the UN's carbon offsetting program Reducing Emissions from Deforestation and Forest Degradation [REDD+] program and the German-led International Sustainability and Carbon Certification [ISCC]); develop renewable energy (the European Union's Renewable Energy Directive [EU-RED]); certify products (the EU-led Roundtable on Sustainable Biomaterials [RSB], Roundtable on Sustainable Palm Oil [RSPO], and Round Table on Responsible Soy [RTRS]); and pay for ecosystem services (Mexico's Payment for Environmental Services-Hydrological [PSA-H in Spanish]). In many countries, these initiatives have led to greater levels of governance by international actors (Bailis and Baka 2011; Moser, Hildebrandt, and Bailis, 2014).

Oil palm producers in Mexico can be certified under international standards of the Roundtable on Sustainable Palm Oil (RSPO) and International Sustainability and Carbon Certification (ISCC). Among the eight RSPO principles and criteria (P&C) are social, environmental, and economic factors that certified growers must address, including committing to transparency; complying with regulations and laws; committing to a viable long-term financial plan; using industry best practices; conserving natural resources and acting environmentally responsibly; considering communities and employees affected by oil palm production; responsibly growing the crop; and committing to on-going improvement in the production process (RSPO, 2017a). Ten companies operating in Mexico hold voluntary RSPO certificates (RSPO, 2017b). The number of RSPO certificate holders in Mexico has risen since 2014, indicating either that more companies are now operating in the country or that existing companies have been pressured to become certified (Byerlee, Falcon, and Naylor, 2017). Two companies operating in Mexico agreed to comply with the six voluntary principles (including the 91 criteria) of the ISCC, which address GHG emissions, environmental impacts, and several aspects related to socio-economic issues (ISCC, 2017).

Mexico lags behind other countries in the number of certification programs it has adopted at the national scale and has only weakly enforced them (Solomon et al., 2015). The few certification programs the country does loosely adhere to are voluntary, and there are many issues on the ground with implementing sustainable practices (Skutsch et al., 2011). Furthermore, certification is only available to organizations that can afford the cost of becoming certified, so it is atypical for smallholder producers to be certified (Byerlee, Falcon, and Naylor, 2017).

Civil society's involvement in sustainable oil palm production

In addition to the top-down and third-party governance of oil palm production, there are local initiatives from civil society. Civil society is comprised of non-governmental, voluntary associations, communities, churches, and other groups of people who freely make innovative and creative decisions without state involvement (Hoffmann, 2011; Seligman, 1992). Civil society can help promote sustainable development, foster linkages between groups of people, and organize resistance against the government (Hoffmann, 2011). Groups acting within civil society can provide an example for and put pressure on governments to adopt similar programs to those devised locally (Hoffmann, 2011).

As far as civil society is involved in guaranteeing the sustainability of oil palm production, the Mexican Federation of Oil Palm (FEMEXPALMA, the Spanish acronym) represents the interests of oil palm producers and extractors (FEMEXPALMA, 2017). The federation works in Chiapas, Campeche, Tabasco, and Veracruz states and charges dues to its members (FEMEXPALMA, 2017). The group's website states that it will "interpret" the RSPO's P&Cs, train and build the capacity of its members, and take an inventory of protected areas, among other projects (FEMEXPALMA, 2017).

Historically, centralized powers in Mexico have remained unaccountable to the people; for example, Presidents Fox and Calderon governed with guidance by and in consultation with corporate interests rather than that of Congress (Jordana, 2010). Mexico has, however, connected international organizations to local NGOs and citizens to implement forestry and agricultural programs (Balderas Torres et al., 2013; Rantala, Hajjar, and Skutsch, 2014). Since the early 2000s, the country has created more legitimacy in regulatory agencies, but not in ways that create more engagement with civil society (Jordana, 2010). Citizens have historically viewed any attempt at holding the government accountable as ineffective, and Mexico's structure of governance and its opaque governance process limits and restricts citizen participation (Hevia, Vergara-Lope, and Avila Landa, 2011; Sabet, 2008). Furthermore, many civil society organizations' efforts toward a bottom-up approach to policymaking have failed (Hevia, Vergara-Lope, and Avila Landa, 2011).

The case of Tabasco, Mexico's oil palm plantations

In 2015, my colleagues and I interviewed 87 people in rural communities in the Teapa and Emiliano Zapata *municipios* of Tabasco, where oil palm plantations are scattered across the landscape. We asked about interviewees' perceptions of their community and employment opportunities; perceptions of the state of and changes in the environment; and their experience with and perceptions of oil palm production. Among participants, the environmental impacts of oil palm production were a central concern.

The rural communities that lie on the outskirts the Teapa city limits are dominated by flat land that is cultivated with banana and oil palm plantations which are interspersed with cattle pastures and secondary forests. The low-lying, flat land is prone to flooding, so many subsistence crops do not grow there. About two hours' driving distance from Teapa, the landscape changes from flat pasture to hilly ranch lands with high altitude mountains and rocky, tree-covered outcroppings. Closer to the communities of human settlement, the landscape changes to rolling hills, some planted in high-value teak and cedar trees while others have been planted in oil palm. The majority of land in Tabasco is under *ejidal* control; the larger private oil palm plantations lie outside the communities and are often owned by absentee landlords or multinational companies.

People frequently discussed the environmental impacts of oil palm. Some told us that they believe the plant requires a lot of water to grow and sucks the local wells and water sources dry. They said that because there is an incentive to grow oil palm, the swamps in the area have dried up. Others said that the plants benefit from areas like Tabasco, where it floods frequently, because they can withstand the water and continue living in it without requiring drainage (in contrast to bananas, which require effective drainage systems). One of the people we interviewed mentioned that the water that runs off the plantations is somewhat contaminated. Another interviewee in the area also told us that their plot is adjacent to an oil palm plantation and that along the border of the two properties the roots of the oil palm have crept over onto his land. Several people described how the roots are very dense along the surface, so it makes it difficult to grow anything else near the oil palm.

In this region, oil palm was planted about 20 years ago under the Agricultural Alliance, a government program that subsidized the production of a variety of crops, including oil palm, as a way to diversify and support the rural economy (Diario Oficial de la Federación, 1998). In one rural community, participants described how a Costa Rican taught the locals how to convert the cattle pastures to oil palm plantations. People told us that growing oil palm requires less work to grow the plants than other crops; however, because the fruit can be harvested every 15 days and requires manual labor to cut the fruit bunches, workers can earn a daily wage working with or in oil palm plantations that is twice as much as they earn working on banana plantations. Many plantations employ local people or migrants from other states. One interviewee mentioned that women are not trained for work in the oil palm plantations, but they can be paid to collect bags of loose oil palm seeds.

In a larger town that is no longer under *ejidal* control, an interviewee told us that people raise cattle instead of planting oil palm because the cattle can quickly be fattened and sold and the land can be used for multiple purposes. This is in contrast to oil palm, which requires a lot of disposable land and requires producers to produce it in bulk in order to sell it. It is a slow-growing plant, we were told, so no one profits immediately from it. In one location, people said they had begun tearing out their oil palm plants because of the low price they receive for the seeds at the market.

Despite information from governmental websites about the availability of oil palm plants for Tabascans (SEDAFOP, 2017), some people we interviewed reported that there are currently no federal subsidies for growing oil palm. When the oil palm first came to the region (via government subsidies), they said, the community was given plants and fertilizer, but now there is very little support of this kind. People also discussed the difficulty producers have in getting bank loans. Besides the cooperative members who were attempting to become RSPO-certified, people we spoke to were not familiar with certification.

On a stretch of semi-paved road through an area of Tabasco with about 600 inhabitants (the asphalt has crumbled off in many sections, possibly due to the high level of use by heavy tractor trailer trucks to and from the oil palm refinery), there is an oil palm cooperative of about 30 small-scale producers. There are about 200 hectares of oil palm planted in this area among the cattle ranches that make up the majority of the land use in the region. The town is more prosperous than others we visited, with several two-story houses and larger houses locked behind sculpted cement walls. However, we were told the town is unsafe at night, with reports of people who hide in a plot of oil palm in between two communities and attack people who pass by.

The cooperative started about 15 years ago. Cooperative members use agroecology farming techniques, inter-planting corn, sweet potato, yucca, and bananas with oil palm during the first three years of establishing the crops. We were told some of the produce is exported to China. A federal government mandate stipulated a minimum requirement of 10 hectares of land in order to receive government subsidies, something which one member labels a discriminatory practice because the people who needed the most support were those who had the small amounts of land and limited access to the federal subsidy programs. This is one of the reasons the cooperative exists, to help people pool their limited resources so they can take advantage of governmental programs. The largest producer has 20 hectares, and the smallest has one hectare of oil palm planted.

When we visited the cooperative, it was in the process of getting RSPO certification. The members work hard to weigh and deliver their fruit to the processing plant within 24 hours of cutting it from the plant, which ensures that it contains the highest oil content; after 24 hours, it starts to rot and loses its quality. Thus, producers need to be close to the plants in order to guarantee the best quality when they sell freshly cut palm fruit. The cooperative sells their product at international market rates to an oil palm processing plant in a nearby *ejido* (communally managed land). The oil palm business has generated employment in the area and the only thing the town lacks, one member said, is enough manual laborers that would allow them to increase their oil palm production.

In an area near Teapa, there are several communities that participate in another cooperative, which grows oil palm, ornamental flowers, and some produce. This cooperative is smaller but operates in a similar manner as the cooperative in Tabasco. Members here also hope to expand their operations and get an RSPO certification, but political and financial barriers have prevented them from doing so. Several landowners who grow oil palm who are not part of the cooperative collaborate more informally, taking their seeds to be weighed and sold with the rest of those from town, and everyone gets paid individually.

Discussion

If Mexico's policies that promote oil palm production (which include poverty reduction and agricultural diversification goals) were sufficient to ensure sustainability, there would be no need for third-party certification or international sustainability standards. However, the use of such practices is telling: the multinational companies that have been certified and the consumers who buy certified products have shown that certification standards are necessary. Comparing the third-party certification and international sustainability standards to Mexico's oil-palm-related policies, with sustainable oil palm production as the goal to achieve, it is clear that domestic policies are far from meeting their goals.

The National Development Plan acknowledges that the government must coordinate various government cabinets to reduce poverty in the country, and to increase productivity the federal government is tasked with assisting local governments in meeting their goals (Gobierno de la Republica, 2013). It is clear that, at the end of the period outlined in the Plan, which spans from 2013–2018, despite the federal government's best intentions to support smallholders, people living in the rural communities feel that the government has not lived up to its promises. Multinational oil palm producers seem to have benefited most because they have access to the larger land holdings required to access government support and can afford to go through the third-party certification processes that make their products appealing to international markets.

Perhaps the problem lies in the disjointed, uncoordinated production of oil palm. Malaysia, a country with a long history of growing the crop for international and domestic markets, has a Palm Oil Board, which in conjunction with the Ministry of International Trade creates oil palm policies (Byerlee, Falcon, and Naylor, 2017). In contrast, Mexico has many policies that point to different goals (poverty reduction, crop diversification, GHG mitigation, export crop production) across various sectors (energy, agriculture), but none of the policies seem particularly effective at meeting their specified aims. In the absence of strong centralized management of oil palm policies and coordination of federal and state level policies, sustainable oil palm production cannot be realized.

Oil palm companies operating in Mexico hold 12 sustainability certificates (ten RSPO and two ISCC certificates). However, this has not led to the widespread production of sustainable oil palm on the ground. It is unclear how much control the Mexican government has over any one company's production process, especially the multinational companies that operate in its territory. However, examining domestic policies, it is clear that the country's administrative priorities do not meet the high sustainability criteria outlined by the international community in the form of certification standards. When there is a mismatch between a country's policies and a third party's motivations, conflict on the ground may occur, and policy goals may not be met (Lambin et al., 2014).

Despite the challenges involved, Mexican civil society could play a role in facilitating and encouraging sustainable oil palm development. FEMEXPALMA has made some headway toward including oil palm producers and extractors in the industry and striving to meet RSPO standards. Currently, smallholders who live on *ejidal* land are able to benefit from oil palm production, by either working on a large plantation or growing it, especially if they share the investment costs and risks through a local cooperative or organization like FEMEXPALMA. As with other agricultural diversification projects in the past (cf. the Agricultural Alliance), rural Tabascans have been able to establish oil palm on their land through government subsidies. However, oil palm production cannot be considered sustainable if the costs are too high to allow people to participate in the industry and if they must be dependent on government subsidies in order to make a living. Unfortunately, despite its potential for moving the country toward sustainably grown oil palm, certified production of the crop in Mexico has not become mainstream. Tabascan smallholder oil palm cooperative members discussed the political and financial difficulties in getting their crops certified despite the length of time they have grown the crop; beyond the difficulties for experienced palm growers in obtaining certification, many rural citizens had never heard of oil palm certification. This indicates that oil palm certification and sustainably grown oil palm is a long way from being realized in the state.

Even without the coordination of oil palm policies in Mexico, land use in the countryside has been increasingly resilient and sustainable in terms of environmental impacts; oil palm plantations have been established on former cattle ranches rather than replacing forests, and smallholders often practice agroforestry, with oil palm often intermingled with other crops. Small-scale land use can be beneficial to the local environment because it can be contained in marginal land to help protect biodiversity and be beneficial to the economy through job creation and income generation in local communities (Fargione, Plevin, and Hill, 2010; Milder et al., 2008).

Possibilities for sustainable oil palm production in Mexico

For implementation of sustainable oil-palm-related policies to occur, several steps would have to be taken by the federal government. Because of Mexico's federated yet concentrated political system in which the federal government is the centralized power, it would have to incentivize cooperation between levels of government, expanding local initiatives to higher levels of government and providing funding to roll out such programs (Valenzuela, 2014). Independent regulatory agencies need to be created; such measures would hold the President accountable, make government more transparent, and promote citizens' interests (Jordana, 2010). The federal government could also incentivize cooperation and coordination at the subnational level to overcome the ineffectiveness of federal policies (Valenzuela, 2014). However, the Mexican government is not able to fully implement and enforce such changes on its own. Structural changes incentivized by third-party actors can help lead the country toward a large-scale sustainable production of oil palm (Selin and VanDeveer, 2009). The government may need to work more closely with the international community to allow for greater participation in certification programs. Pressure from potential consumers or investors to ensure that companies comply with sustainability criteria may also be necessary to promote a resilient oil palm economy in Mexico (Byerlee, Falcon, and Naylor, 2017; Heberlein, 2012).

Finally, the idea that subnational policies can trickle up to higher levels of government may also apply to leadership (Cole, 2015). Despite the stronghold that sectors of government have on the policy process, it would be in Mexico's interest to include civil society participation in the process. More public participation either through stronger democratic channels or by fostering civil society development and trust in government would be beneficial to the government beyond the issue of producing oil palm (Cole, 2015; Hoffmann, 2011). Programs that fit the local context and environment are especially important for achieving local buy-in and ensuring participation locally (Cole, 2015). FEMEXPALMA could work with state or regional politicians and community organizations to coordinate national level policies for sustainable oil palm production.

Conclusion

A hybrid government–governance mix for promoting sustainability and resilience in the development of oil palm production in Mexico, rather than pure governance by private actors or dependence solely on the Mexican government, is realistic, but not until Mexico allows multiple actors, including civil society organizations, to engage in the process of governing their resources via policies (Ponte and Daugbjerg, 2014). The country's federal policies need to be strengthened to incorporate regional differences and needs into policy (Hunsberger et al., 2014; Moser, Hildebrandt, and Bailis, 2014; Stattman and Mol, 2014). There needs to be more transparency and accountability on the part of the Mexican government, international community, and global companies to follow sustainability criteria to produce responsible products that enable the rural poor to control their livelihoods better. Local producers should be part of the discussion about what they expect and need from these production systems (Moser, Hildebrandt, and Bailis, 2014). If the federal government recognizes and supports initiatives like those of FEMEXPALMA and local cooperatives, those organizations can fulfill the government's task of delivering context-specific benefits to rural communities. This would alleviate some of the government's burden of coordinating what is happening in local communities and help the country move toward achieving its dual goals for poverty alleviation and agricultural diversification. Third-party certification organizations can then liaise with the federal government and organizations to ensure that oil palm development meets international standards and provides social, economic, and environmental benefits for the rural communities.

Acknowledgements: this work was funded by the National Science Foundation's Partnerships in International Research and Education (PIRE) Program.

References

Alemán-Nava, G., Meneses-Jácome, A., Cárdenas-Chávez, D., Díaz-Chavez, R., Scarlat, N., Dallemand, J., Ornelas-Soto, N., García-Arrazola, R., and Parra, R. (2014). Bioenergy in Mexico: Status and perspective. *Biofuels, Bioproducts and Biorefining*, 9(1), pp. 8–20.

Bailis, R. and Baka, J. (2011). Constructing sustainable biofuels: Governance of the emerging biofuel economy. *Annals of the Association of American Geographers*, 101(4), pp. 827–838.

Balderas Torres, A., Ontiveros Enríquez, R., Skutsch, M., and Lovett, J. (2013). Potential for climate change mitigation in degraded forests: A study from La Primavera, México. *Forests*, 4(4), pp. 1032–1054.

Byerlee, D., Falcon, W., and Naylor, R. (2017). *The Tropical Oil Crop Revolution: Food, Feed, Fuel, and Forests*. Oxford: Oxford University Press.

Cámara de Diputados (2008). Le y de promoción y desarrollo de los bioenergéticos. Available at: www.diputados.gob.mx/LeyesBiblio/pdf/LPDB.pdf [Accessed 14 May 2017].

Cole, D. (2015). Advantages of a polycentric approach to climate change policy. *Nature Climate Change*, 5(2), pp. 114–118.

Diario Oficial de la Federación (1998). Programa de Alianza para el Campo. Available at: dof.gob.mx/nota_detalle.php?codigo=4881535&fecha=03/06/1998 [Accessed 5 January 2016].

Diario Oficial de la Federación (2008). Ley de promoción y desarrollo de los bioenergéticos. Available at: dof.gob.mx/nota_detalle.php?codigo=5029330&fecha=01/02/2008 [Accessed 5 May 2016].

Eastmond A., García, C., Fuentes, A., and Becerril-García, J. (2014). Mexico. In: (eds.) Solomon, B. and Bailis, R., *Sustainable Development of Biofuels in Latin America and the Caribbean*, 1st ed. New York: Springer, pp. 203–222.

Fargione, J., Plevin, R., and Hill, J. (2010). The ecological impact of biofuels. *Annual Review of Ecology, Evolution, and Systematics*, 41, pp. 351–377.

Federacion Mexicana de Palma de Aceite (FEMEXPALMA) (2017). Inicio. Available at: www.femexpalma.com.mx/ [Accessed 17 July 2017].

Gay, C. and Estrada, E. (n.d.). Programa de cambio y variabilidad climaticos: The Clean Development Mechanism under a Mexican perspective. Available at: www.atmosfera.unam.mx/cclimat/documents/cambio_climatico/articulos_discusion/cdm_mexico_dos-1.pdf [Accessed 12 December 2015].

Gobierno de Mexico (2012). Mexico has a potential of more than two million hectares for oil palm cultivation: INIFAP. Available at: www.sagarpa.gob.mx/saladeprensa/2012/Paginas/2015B347.aspx [Accessed 27 July 2017].

Gobierno de la Republica (2013). Plan Nacional de Desarrollo 2013–2018. Available at: pnd.gob.mx/ [Accessed 20 May 2017].

Heberlein, T. (2012). *Navigating Environmental Attitudes*. Oxford: Oxford University Press.

Hevia, F., Vergara-Lope, S., and Avila Landa, H. (2011). Participación Ciudadana en México: Consejos Consultivos e Instancias Públicas de Deliberación en el Gobierno Federal. *Perfiles Latinoamericanos*, 18(38), pp. 65–88.

Hoffmann, M. (2011). *Climate Governance at the Crossroads: Experimenting with a Global Response After Kyoto*. Oxford University Press.

Hooghe, L. and Marks, G. (2003). Unraveling the central state, but how? Types of multi-level governance. *American Political Science Review*, 97(2), pp. 233–243.

Howlett, M. (2000). Managing the "hollow state": Procedural policy instruments and modern governance. *Canadian Public Administration*, 43(4), pp. 412–431.

Howlett, M., Ramesh, M., and Perl, A. (2009). *Studying Public Policy: Policy Cycles and Policy Subsystems*, 3rd ed. Toronto: Oxford University Press.

Hunsberger, C., Bolwig, S., Corbera, E., and Creutzig, F. (2014). Livelihood impacts of biofuel crop production: Implications for governance. *Geoforum*, 54, pp. 248–260.

International Sustainability and Carbon Certification (ISCC) (2017). All certificates. Available at: www.iscc-system.org/certificates/all-certificates/ [Accessed 27 July 2017].

Jordana, J. (2010). Autonomous regulatory agencies in democratic Mexico. *Law & Business Review of the Americas*, 16, pp. 753–779.

Lambin, E., Meyfroidt, P., Rueda, X., Blackman, A., Börner, J., Cerutti, P., Dietsch, T., Jungmann, L., Lamarque, P., Lister, J., and Walker, N. (2014). Effectiveness and synergies of policy instruments for land use governance in tropical regions. *Global Environmental Change*, 28, pp. 129–140.

Milder, J., McNeely, J., Shames, S., and Scherr, S. (2008). Biofuels and ecoagriculture: Can bioenergy production enhance landscape-scale ecosystem conservation and rural livelihoods? *International Journal of Agricultural Sustainability*, 6(2), pp. 105–121.

Moser, C., Hildebrandt, T., and Bailis, R. (2014). International sustainability standards and certification. In: Solomon, B. and Bailis, R., eds., *Sustainable Development of Biofuels in Latin America and the Caribbean*, 1st ed. New York: Springer, pp. 27–69.

Ponte, S. and Daugbjerg, C. (2014). Biofuel sustainability and the formation of transnational hybrid governance. *Environmental Politics*, 24(1).

Rantala, S., Hajjar, R., and Skutsch, M. (2014). Multilevel governance for forests and climate change: Learning from southern Mexico. *Forests*, 5(12), pp. 3147–3168.

Rodríguez, O., Vázquez, A., and Gamboa, C. (2014). Drivers and consequences of the first Jatropha curcas plantations in Mexico. *Sustainability*, 6(6), pp. 3732–3746.

Romero-Hernández O., Masera O., Romero S., and Grunstein, M. (2011). Legal and institutional frameworks at national and subnational levels for biofuel promotion in Mexico, Working Paper, 63, CIFOR, Bogor: Indonesia.

Roundtable on Responsible Palm Oil (RSPO) (2017a). Principles and criteria for the production of sustainable palm oil. Available at: www.rspo.org/publications/download/4b4296c7bb85cb3 [Accessed 27 July 2017].

Roundtable on Responsible Palm Oil (RSPO) (2017b). Supply chain certificate holders. Available at: www.rspo.org/certification/supply-chain-certificate-holders?keywords=&country=Mexico [Accessed 27 July 2017].

Rubio, J. (2017). Palma de aceite: El nuevo "oro verde". Available at: novedadesdetabasco.com.mx/2017/02/23/palma-de-aceite-el-nuevo-oro-verde/ [Accessed 25 May 2017].

Sabet, D. (2008). Thickening civil society: Explaining the development of associational life in Mexico. *Democratisation*, 15(2), pp. 410–432.

Secretaría de Desarrollo Agropecuario, Forestal y Pesquero (SEDAFOP) (2017). Impulsa Sedafop cultivo de palma de aceite. Available at: sedafop.tabasco.gob.mx/content/impulsa-sedafop-cultivo-de-palma-de-aceite [Accessed 25 May 2017].

Seligman, A. (1992). *Idea of Civil Society*. London: Simon and Schuster.

Selin, H. and VanDeveer, S., (2009). Climate leadership in northeast North America. In: Selin H., and VanDeveer, S., eds., *Changing Climates in North American Politics: Institutions, Policymaking, and Multilevel Governance*, 1st ed. Cambridge: MIT Press, pp. 111–136.

Skutsch, M., de los Rios, E., Solis, S., Riegelhaupt, E., Hinojosa, D., Gerfert, S., Gao, Y., and Masera, O. (2011). Jatropha in Mexico: Environmental and social impacts of an incipient biofuel program. *Ecology and Society*, 16(4), p. 11.

Solomon, B., Banerjee, A., Acevedo, A., Halvorsen, K., and Eastmond, A. (2015). Policies for the sustainable development of biofuels in the Pan American Region: A review and synthesis of five countries. *Environmental Management*, 56(6), pp. 1276–1294.

Stattman, S. L. and Mol, A. P. (2014). Social sustainability of Brazilian biodiesel: The role of agricultural cooperatives. *Geoforum*, 54, pp. 282–294.

Tollefson, C., Gale, F., and Haley, D. (2008). *Setting the Standard: Certification, Governance, and the Forest Stewardship Council.* Vancouver: UBC Press.

United Nations Framework Convention on Climate Change (UNFCCC) (2016). Mexico: Intentional Nationally Determined Contribution. Available at: www4.unfccc.int/submissions/INDC/Published%20Documents/Mexico/1/MEXICO%20INDC%2003.30.2015.pdf [Accessed 15 April 2016].

United States Department of Agriculture Foreign Agricultural Service (USDA-FAS) (2017). Crop Explorer. Available at: www.pecad.fas.usda.gov/cropexplorer/cropview/commodityView.aspx?cropid=4243000&sel_year=2017&rankby=Production [Accessed 30 July 2017].

Valenzuela, J. (2014). Climate change agenda at the subnational level in Mexico: Policy coordination or policy competition? *Environmental Policy and Governance*, 24(3), pp. 188–203.

CONCLUSION

Environmental policy and the pursuit of just sustainability

Chelsea Schelly and Aparajita Banerjee

Human decision-making has profound effects on the ability of the planet to sustain our species and all others. This is true for our individual consumption choices and especially true for professional environmental regulators, planners, managers, and policy makers who make and implement choices regarding how humans utilize resources from the environment to meet their physically and socially defined needs and desires with consequences for water, land, health, and social justice. This volume explores empirical examples of how policy decision-making is navigated and executed and the implications of those decision-making choices for humans and their environments. The case studies suggest the importance of four main themes for environmental policy in pursuit of sustainability: 1) issues of scale mismatch, geographically and geopolitically; 2) issues of policy silos, wherein environmental policy is pursued as if divorced from economic and social policy issues; 3) issues of social justice, wherein environmental policy is pursued without just consideration of impacts; and 4) issues of leadership, whereby given the complexities of geographical and geopolitical scale incongruences, existing relations of power and the balancing of water, land, health, and social justice considerations, there is a lack of clear leadership to manage the uncertainties of environmental change.

Issues of scale mismatch

Contemporary environmental problems are complex, wicked problems (Balint, Stewart, and Desai, 2011). This complexity is evident in the ways environmental issues cross geographical scales such that upstream activities become downstream hazards and localized pollution issues can become regionally, nationally, or even globally impactful. The geographical complexities involved in how environmental problems transverse geographical scales is further complicated by the mismatch between geographically relevant scales for thinking about environmental issues and the geopolitical jurisdictional scales in which policymaking and policy implementation take place.

For example, in Chapter 1, M. Dawn King demonstrates that non-point source pollution is a geographically deviant environmental issue, as pollution sources cannot easily be pinpointed to a single source but can have widespread impacts across watersheds (crossing community and even state lines), and thus management of such pollution becomes a difficult process of accounting and regulation. Further, while water management may be most effective at a watershed scale, most political institutions are designed to serve federated jurisdictions including local communities, regions, states, or national level concerns, many of which overlap at the scales most relevant to water management. Learning to work across and even beyond these traditional geopolitical scales in order to address the most relevant scales for managing and developing solutions for environmental problems may require attentiveness to the ineffectiveness of current scales of policy decision-making for environmental issues that cross geographical and geopolitical boundaries. These same issues are again highlighted in Chapter 3, with the Hampton Roads Intergovernmental Pilot Project providing a real life example of the complexities of managing water systems across geographical scales, geopolitical jurisdictions, and the uncertainties of long term change over temporal scales not typically addressed in traditional policymaking.

The issues of geographical and geopolitical scale mismatch are also raised in Chapter 2, in discussion of the controversies of water management policy in the Upper Klamath Basin, and in Chapter 4's discussion of public lands management. Both cases demonstrate how environmental policy impacts a complex and multi-scalar network of individuals, local communities, and regions and how there is often (or perhaps always) no straightforward answer of the most effective or appropriate scales at which to develop environmental policy given its complex and multi-scalar impacts. However, both chapters also demonstrate that while there may be effectiveness in local mobilization and appropriateness in involving local perspectives in policy, federal agencies and actors still have immense power to shape the environmental policy landscape (albeit power that may be most impactful in its ability to hinder effective change, as in Chapter 2, rather than power in proactively creating change, a power that can itself be curbed by regional and local policy actions). Local groups and state policies certainly have increasing power to effect change in times of weak national scale leadership on sustainability issues, as is evidenced by the proliferation of state level renewable portfolio standard (RPS) requirements for the United States' electric utilities in the absence of a national level RPS (Carley, 2011; Mormann, 2017). The role of individual US states in environmental protection policy has evolved over time (Rabe, 2016) and is likely to continue to do so; however, these chapters demonstrate that federal government powers are still an essential consideration in the geographical and geopolitical complexities of enacting environmental policy.

These issues of the complexity of geopolitical scales in creating effective environmental policy and effectively managing environmental change are a continuous theme throughout the volume, including how solid waste governance in Mexico is managed by multiple institutions across multiple political scales and how this geopolitical scale complexity may contribute to ineffective waste management (Chapter 6), the scalar complexities of managing access to resources on privately owned land via federal regulations and very limited localized power to regulate resource development (Chapter 8), the complexities involved in local implementation of federal policy (Chapter 9), and the challenges of managing for environmental change across geographical, geopolitical, and temporal scales as well as agency jurisdictions and mandates (Chapter 11). There is certainly no evidence that there is one best scale at which to develop environmental policy; instead, the evidence demonstrates that geopolitical jurisdictions of policy formation and implementation may be most effective when they are effectively integrated across multiple scales, just like the ecosystems they are intended to manage.

Issues of policy silos

Environmental policymaking also seems to often occur in a policy silo, where environmental issues are considered in isolation from one another and from the larger context of how environmental policy is also relevant for economic policy, development policy, and social policy, among many other policy categories and impacts. As demonstrated in Chapter 1, even water management policy (to address non-point source pollution) has been developed without consideration for climate policy (to address climate-changing emissions), and more effective solutions may be found by stepping outside these isolated policy silos to understand and even develop policy that best serves to promote ideal conditions for water, land, air, climate, economies, communities, health, and justice – as life itself is complex and multi-faceted, policymaking must also contend with the complexities of policy goals and outcomes.

Chapter 1 also demonstrates that environmental policy is, in many ways, also economic policy; decision-making about regulating environmental impacts often involves subsidies or other monetary incentives in which some individuals, technologies, businesses, and industries benefit while others are impeded. Chapter 2 demonstrates this clearly, as water management controversy sometimes boils down to a controversy over whose farm may have enough water to turn a profit, who may get to eat fish for dinner, or how much you pay for the electricity that turns your lights on. In Chapter 4, as well, designation and management of public lands also involve clear implications for who gets to benefit economically from the grazing, extraction, or tourism activities that take place on or in conjunction with public lands. Chapter 10 demonstrates that policy silos also shape assumptions about the motivations of entities regulated by environmental policy in ways that may not be consistent with effectively promoting holistically beneficial practices for the environmental, economic, health, and social outcomes resulting from the activities of a regulated entity.

In Chapter 6, the relationship between environmental and economic policy is clear in the lack of regulation shaping the production end of the waste stream (as in product packaging as well as planned obsolescence), as is the relationship between environmental policy and outcomes for human and environmental health, a theme raised again and again in Chapter 7, Chapter 8, Chapter 9, and Chapter 12. Environmental policy effectively involves decisions about access to clean air, clean water, and land suitable for human habitation, services that are essential for human life but are nonetheless not equitably distributed among human beings. The human health impacts are an essential consideration for any environmental policymaking process, but policy silo effects often create difficulties for integrated thinking, planning, and policymaking across all the various dimensions involved in quality of life.

In order to make effective environmental policy, it is essential to think about the intersecting nexus of impacts connecting air and water, water and land, land and climate, etc. It is also arguably essential to recognize that environmental policy always has real implications for economies, community development, and actual human health. Perhaps thinking through the potentially unintended impacts of any policy decision (Merton, 1967; Schelly, 2014) in terms of its widespread impacts across the human experience is an ideal first step in any policy process. Books, courses, departments, and agencies focused on environmental policy (Rosenbaum, 2013; Vig and Kraft, 2016; and of course the current volume!) may all suggest an artificial separation between environmental issues and other issues of real human life, indicating that environmental policy is somehow separate from other policy, management, or moral issues. This false suggestion may work to hinder progress toward training policy experts, advocates, and scholars who recognize the integrated impacts of environmental decision-making on economies, technologies, relations of power, and systems of social justice.

Issues of social justice

In this volume, we aim to especially highlight that environmental policy, in every instance presented here, raises essential questions about the distribution of access to resources, wealth, and well being in ways that characterize issues of social justice. Environmental policymaking takes place within existing contexts of social injustice and can work to either exacerbate or eradicate that injustice, but arguably the most common impact is for environmental policy to make conditions of injustice worse or certainly no better simply because environmental policy explicitly ignores its implication in issues of social justice.

Sometimes, the simplistic language of costs, payments, benefits, or burdens can capture these questions of social justice. In the case of Chapter 1, a clear question exists regarding who should have to pay for the management of non-point source waste in the form of chicken litter: should the costs of management be the responsibility of corporate profit makers or the contracted farmers? The question may be answered via contract negotiation; however, environmental policymaking can also be used to explicitly address questions such as who pays, who benefits, and who bears the burden of environmental policy.

However, issues of social justice go beyond questions of payments and costs to issues of access and inequitable access to both quality of life and decision-making power. In Chapter 5, and also highlighting the complexities of geographical and geopolitical scale in managing resource extraction with potentially monumental impacts on local communities and regional ecosystems via federal rather than local policy, Kristin Olofsson's research demonstrates that policy actors involved in issues of resource extraction via hydraulic fracturing in the realm of federal policy are not commonly exposed to the impacts of hydraulic fracturing on a personal level. This inevitably raises questions about procedural justice (Sovacool et al., 2017) and the rights of individuals and communities to participate in environmental decision-making that impacts them directly and to avoid being subject to decisions made by those who are not at all personally impacted, issues also raised by Chapter 2.

Chapter 7 directly considers the justice implications of environmental policy-making in the United States, particularly through the lack of consequences in clear cases of environmental injustice, where environmental management decisions are made in ways that disproportionately impact already marginalized communities. Environmental injustices such as the water crisis in Flint, Michigan, are devastatingly too numerous to count, both in the United States and around the world as the result of global commodity chains and flows. People who are poor and people of color are overwhelmingly burdened with the environmental hazards and risks associated with productive and consumptive activities, and all environmental policy choices involve choices about the communities and individuals that will face the burden of exposure to unclean water, air, and land as well as diminished human health and quality of life.

The social justice implications of how land is used to extract energy resources is the topic of Chapter 8, which explicitly calls for management approaches that recognize the injustice in expecting fair negotiations of land use contracts given the existing relations of power in society. Corporate entities stand to gain massively from the exploitation of natural resources, with huge consequences for human health; these private, corporate actors leverage a form of persuasive power to insinuate that they are acting with the authority of the sovereign power (as the state is often seen as the legitimate holder of sovereign power) in ways that manipulate the process of contracted access to land for development. Given the existing relations of power in which economic actors often do have more structural and cultural power and in which they use persuasive power to convince land owners that they also have the legitimacy of sovereign power, Chapter 8 argues that it is the responsibility of environmental policy makers to recognize the uneven distribution of power that shapes the terrain of environmental management and to proactively work to develop ways of evening the playing field so that environmental management decisions with enormous consequences for community health and well being can be negotiated fairly and in a balanced way in the face of economic pressures for development.

Chapter 9 also explicitly deals with issues of social justice, specifically how current distributions of socio-political power shape access to resources that are intended to benefit individual health and well being as well as global climate health and well being. In this case, political power and party affiliation influenced the distribution of improved cookstove technology in a way that diminished the program's effectiveness while solidifying the legitimacy of the current distribution of power across both gendered and party lines. Themes of social justice are also considered in Chapter 12, as individuals with the least amount of power in the form of land holdings are also least able to benefit from either federal government programs intended to promote sustainable development or international programs intended to open access to markets geared for sustainable products.

Making and implementing environmental policy across all scales and levels of governance always involves questions of social justice; some people get to participate in decision-making and get to benefit from the intended outcomes while others are left to bear the burdens of decisions they did not make. In these ways, environmental policy always involves grappling with questions of justice, in terms of who has the right to participate in decision-making and who has the right to the benefits of a quality life with clean water, land, and air. Environmental policymaking that does not explicitly address issues of power and justice when addressing implementation is likely to exacerbate current injustices. By simply ignoring injustice, environmental policy often works to perpetuate it.

Issues of lacking leadership

A final theme presented in this volume relates to the lack of clear leadership for making effective environmental policy in pursuit of sustainability. The complexities in identifying and navigating leadership in this terrain include all of the issues raised above: the geographical and geopolitical scales (as well as the temporal scales) are complex, sometimes overlapping and sometimes mismatched entirely; the policy silos that occur around particular issues or sectors limit the ability of emerging leaders working across resource areas or policy realms to be effective in implementing change; and promoting long-term resilience via environmental policy requires grappling with existing relations of power, issues of social injustice, and the role of environmental policy in attenuating or exacerbating these issues. The geopolitical and jurisdictional scales at which leadership could be most effective in creating environmental policy for just sustainability remain unclear, and an overall sense of lacking leadership pervades the challenges of policy success. In Chapter 3, the Hampton Roads Intergovernmental Pilot Project faced the challenge of lacking clear leadership in thinking about the project's success and longevity in the future. In Chapter 8, clear leadership would be required but has yet to effectively emerge in order to combat the challenge of persuasive power via local policy and advocacy channels. In Chapter 11, lack of clear leadership was identified as a challenge to creating long-term plans for managing landscape-scale environmental change.

One particularly effective way for leadership to emerge in the realm of environmental decision-making may involve promoting a sense of efficacy among individuals and groups whose actions are environmentally impactful. Efficacy refers to the sense that one's actions make a difference, and given the multifaceted and global scale of environmental issues, it is challenging to maintain a sense of efficacy in the face of such complexities. Yet research suggests that people who feel a sense of efficacy are much more likely to engage in environmentally responsible behaviors (Schelly et al., 2011; 2012). Effective environmental leadership may require navigating the complexities of geographical and geopolitical scale as well as issue intersections in order to identify areas of truly impactful change and communicate about the possibilities for this change. This may also require speaking truth to power to identify when it is not individual but rather collective corporate choices that are causing the greatest extent of environmental damage (Freudenburg, 2006; Collins, Munoz, and JaJa, 2016; Jorgenson, Longhofer, and Grant, 2016; Greenberg, 2017) and being willing to find innovative ways of utilizing efficacy to promote change in larger societal practices rather than focusing on individual behaviors.

Potential leaders face many challenges in environmental policymaking, including the difficulty of acquiring and applying scientific data at the correct temporal and spatial scales, the complexity of coordinating among land owners and managers across geographical scales and social institutional units (e.g., private land, public land, corporate actors), and the difficulty of making decisions in an environment of climate uncertainty. Political regimes shift and the geopolitical scales at which environmental leadership may be most effective change over time, and environmental policy involves complex choices about water, air, and land but also about technological change, economic development, human health, and social justice. Yet leaders that are willing to navigate the spatially and socially complex world of environmental policy are absolutely essential for pursuing social justice and sustainability through effective policy.

Conclusion: pursuing just sustainability through environmental policy

As of the year 2017, it has been 30 years since the term sustainable development was popularized by the World Commission on Economic Development in 1987 and concepts of sustainability began to circulate widely in socio-political domains. Some of the challenges identified by the different authors in this volume in pursuing sustainability via environmental policy can perhaps be traced back to how sustainability was imagined as an economic term rather than encompassing "essentially ethico-political objectives" (Jacobs, 1995, p. 65). Given the focus on economic development in much of the initial sustainability discourse, and perhaps given more fundamental issues related to relations of power and modes of production, there has rarely been a balance between the social, economic, and environmental dimensions of sustainability and trade-offs mostly favor the economic dimension at the cost of the other two (Lorek and Spangenberg, 2014). This applies to both global governance and policy frames that trickled down to even regional and local levels.

Today, public policies continue to be steered towards meeting economic goals and sustainability has become, according to some, an "empty phrase and at worst a Trojan horse" (Voß and Kemp, 2015, p. 3). Even environmental policies have been, for the most part, aspirational in nature, rather than setting down mandatory guidelines (with few exceptions, like the Endangered Species Act). As a result, social considerations based on human rights, inequality, redistribution, and capabilities are often missing in environmental policymaking. Even environmental concerns have often been sidelined, as living within environmental limits conflicted with continuous economic growth (Lorek and Spangenberg, 2014). Therefore, not surprisingly, 30 years hence, we continue to grapple with the pursuit of sustainability.

The key themes that run throughout the chapters of this book, as discussed at the beginning of this chapter, may provide insight to finding some solutions. As many of the chapters in the book identified, there is a need for innovation in environmental policy processes so that new forms of collaborative problem solving can tackle the non-linear, dynamic, and unpredictable nature of environmental problems. Existing policy silos can be replaced by a policy process that is coherent across sectors in such a way that policy problems are framed and solutions are identified by recognizing their interconnectedness across sectors, with other problems, the scales at which they operate, and direct and indirect impacts of implementable actions. Innovation in the policy realm is also necessary to overcome the path dependency of past and current policies, many of which perhaps undervalue the environmental challenges human society is facing; some of these challenges have never happened before, at least not at present scales. Though breaking out of existing conditions in our socio-political regimes can meet resistance, particularly from those who benefit from existing social structures, the quest for alternatives should involve grappling with making effective environmental policy that is attentive to complexities and issues of scale, comprehensively addresses social issues, is proactive in addressing social justice, and emerges within a vision of comprehensively embracing possibilities for sustainability.

References

Balint, P. J., Stewart, R. E., and Desai, A., 2011. *Wicked Environmental Problems: Managing Uncertainty and Conflict*. Island Press.

Carley, S., 2011. The era of state energy policy innovation: A review of policy instruments. *Review of Policy Research*, *28*(3), pp. 265–294.

Collins, M. B., Munoz, I., and JaJa, J., 2016. Linking 'toxic outliers' to environmental justice communities. *Environmental Research Letters*, *11*(1), p. 15004.

Freudenburg, W. R., 2006. Environmental degradation, disproportionality, and the double diversion: Reaching out, reaching ahead, and reaching beyond. *Rural Sociology*, *71*(1), pp. 3–32.

Greenberg, P., 2017. Disproportionality and resource-based environmental inequality: An analysis of neighborhood proximity to coal impoundments in Appalachia. *Rural Sociology*, *82*(1), pp. 149–178.

Jacobs, M., 1995. Sustainable development, capital substitution and economic humility: A response to Beckerman. *Environmental Values*, *4*(1), pp. 57–68.

Jorgenson, A., Longhofer, W., and Grant, D., 2016. Disproportionality in power plants' carbon emissions: A cross-national study. *Scientific Reports*, 6.

Lorek, S. and Spangenberg, J. H., 2014. Sustainable consumption within a sustainable economy – beyond green growth and green economies. *Journal of Cleaner Production*, *63*, pp. 33–44.

Merton, R., 1967. Manifest and Latent Functions. Chapter 3 in *On Theoretical Sociology*. New York: Free Press.

Mormann, F., 2017. Constitutional challenges and regulatory opportunities for state climate policy innovation. *Harvard Environmental Law Review, 41*, pp. 189–242.

Rabe, B., 2016. Racing to the top, the bottom, or the middle of the pack? The evolving state government role in environmental protection. In *Environmental Policy: New Directions for the Twenty-First Century*, Norman J. Vig and Michael E. Kraft (eds.). Washington DC: CQ/Sage, pp. 33–57.

Rosenbaum, W. A., 2013. *Environmental Politics and Policy*. CQ Press.

Schelly, C., 2014. Implementing renewable energy portfolio standards: The good, the bad, and the ugly in a two state comparison. *Energy Policy, 67*, pp. 543–551.

Schelly, C., Cross, J. E., Franzen, W. S., Hall, P., and Reeve, S., 2011. Reducing energy consumption and creating a conservation culture in organizations: A case study of one public school district. *Environment and Behavior, 43*(3), pp. 316–343.

Schelly, C., Cross, J. E., Franzen, W., Hall, P., and Reeve, S., 2012. How to go green: Creating a conservation culture in a public high school through education, modeling, and communication. *The Journal of Environmental Education, 43*(3), pp. 143–161.

Sovacool, B. K., Burke, M., Baker, L., Kotikalapudi, C. K., and Wlokas, H., 2017. New frontiers and conceptual frameworks for energy justice. *Energy Policy, 105*, pp. 677–691.

Vig, N. J. and Kraft, M. E. eds., 2016. *Environmental Policy: New Directions for the Twenty-First Century*, Ninth Edition. CQ Press.

Voß, J. P. and Kemp, R., 2015. *Sustainability and Reflexive Governance: Introduction*. Technische Universität Berlin.

INDEX